IDAHO'S LARGEST RIVER
takes its name from the Indians
who painted snakes on sticks
to mark territory and frighten
their neighbors. These people,
according to legend, had an
unusual way of greeting
strangers: a sinuous, outward
motion of their hand that
signified "People of the
Snake." In 1811 or 1812, British
and American trappers applied
the name to the Lewis Fork of
the Columbia, the Nile of
Idaho's flat, crescent-shaped
prairie or high desert—
the Snake River Plain.

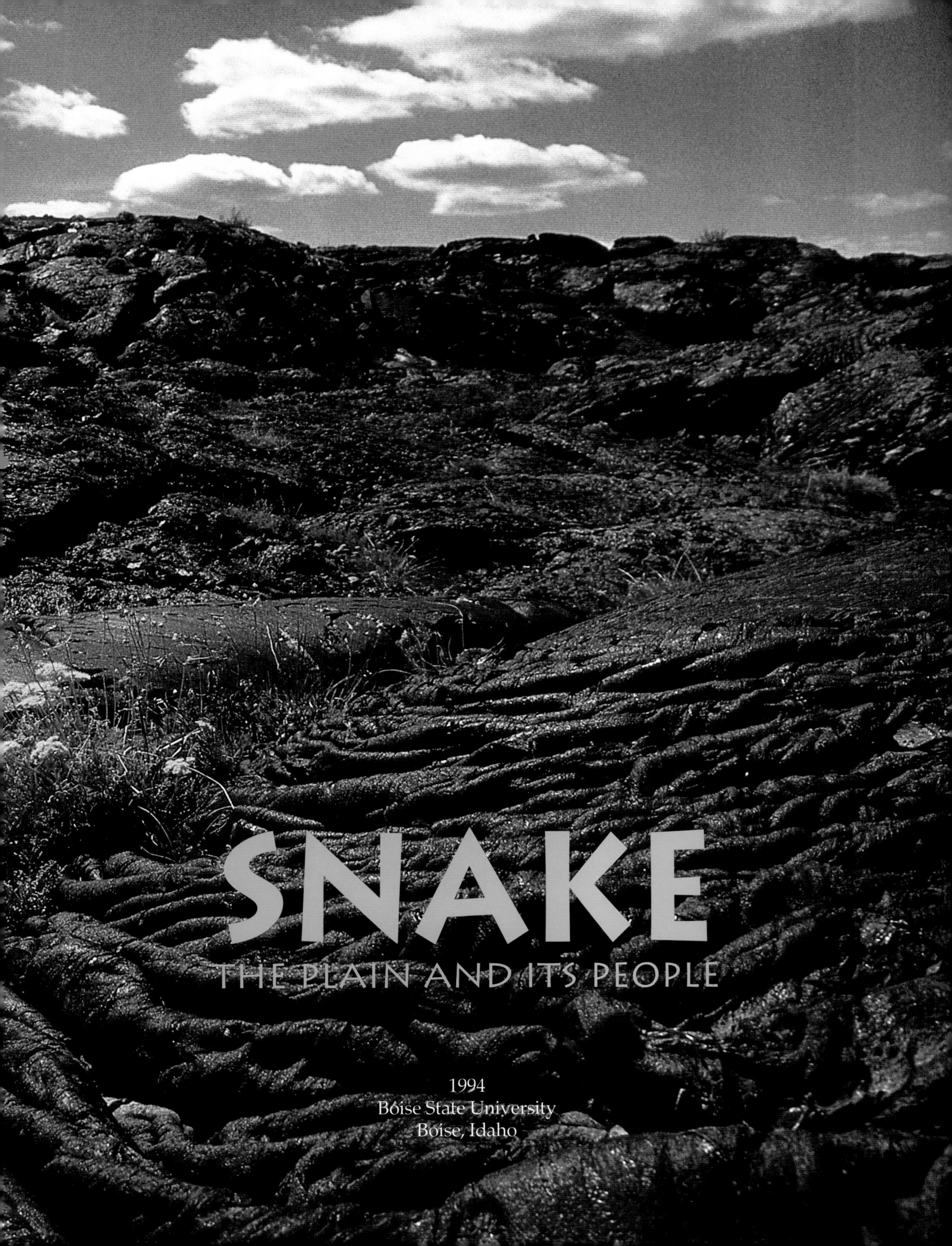

SNAKE
THE PLAIN AND ITS PEOPLE

1994
Boise State University
Boise, Idaho

PHOTOS

p. 1: pictograph at Danskin rock shelter; pp. 2–3: canyon lands, East Fork of the Owyhee River; pp. 4–5: highway through the upper Snake country; pp. 6–7: Craters of the Moon lava flows; pp. 8–9: southern Idaho farmer; pp. 10–11: postcard of lava terraces along the Snake River, about 1942.

© 1994
Boise State University

Ordering information:
Boise State University
University Relations
1910 University Drive
Boise, ID 83725
(208) 385-1577
All Rights Reserved
Library of Congress
Card No. 93-073992
ISBN: 0-932129-12-9

EDITORS

Todd Shallat, editor
Larry Burke, managing editor
Josie Fretwell, graphics editor
Chris Latter, designer

AUTHORS

E.B. Bentley
Bill Bonnichsen
John Freemuth
Bill Hackett
Glenn Oakley
F. Ross Peterson
Mark Plew
Todd Shallat
Steve Stuebner

ADVISORS

Dave Clark
Robert Sims

SPONSORS

Published with the support of
the Craters of the Moon
Natural History Association,
the Boise State University
College of Social Sciences and
Public Affairs, Hemingway
Western Studies Series, Office
of University Relations and the
Idaho Centennial Commission.

DISCLAIMER

This publication was partially funded by the Craters of the Moon Natural History Association, Inc. The contents and opinions contained in this book, however, are those of the authors and do not necessarily reflect the views or policies of the Craters of the Moon Natural History Association, Inc., Craters of the Moon National Monument, the National Park Service or the Department of the Interior.

BISBEE PHOTO

CONTENTS

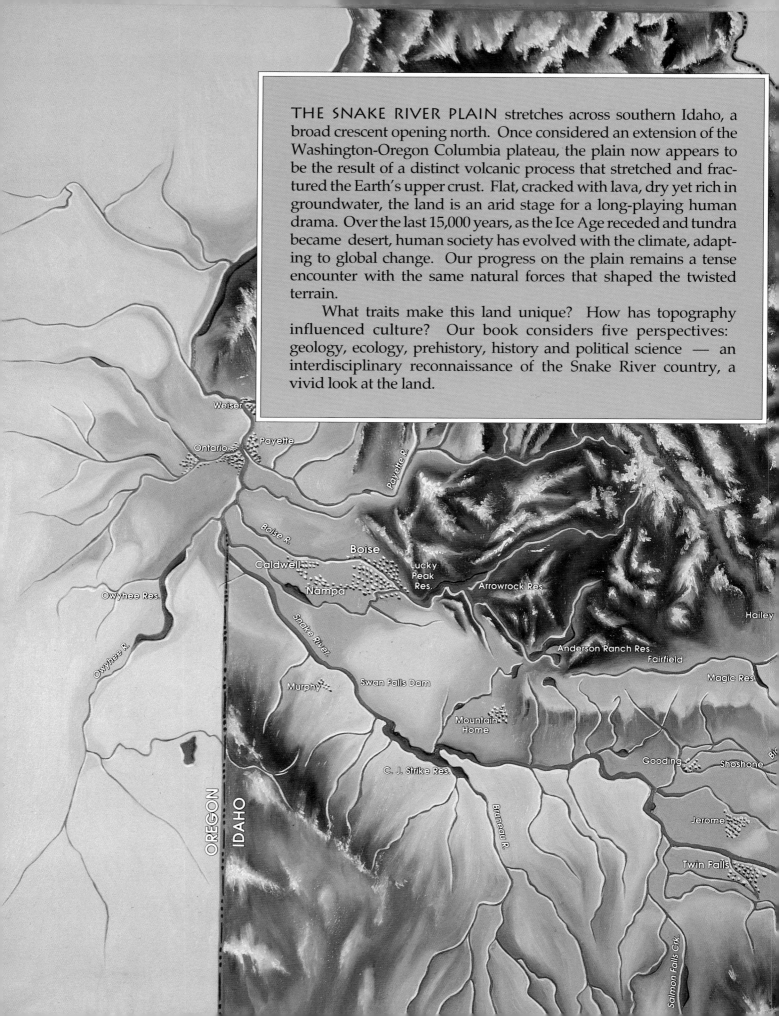

THE SNAKE RIVER PLAIN stretches across southern Idaho, a broad crescent opening north. Once considered an extension of the Washington-Oregon Columbia plateau, the plain now appears to be the result of a distinct volcanic process that stretched and fractured the Earth's upper crust. Flat, cracked with lava, dry yet rich in groundwater, the land is an arid stage for a long-playing human drama. Over the last 15,000 years, as the Ice Age receded and tundra became desert, human society has evolved with the climate, adapting to global change. Our progress on the plain remains a tense encounter with the same natural forces that shaped the twisted terrain.

What traits make this land unique? How has topography influenced culture? Our book considers five perspectives: geology, ecology, prehistory, history and political science — an interdisciplinary reconnaissance of the Snake River country, a vivid look at the land.

LEGEND

NORTH

24 miles

MONTANA

IDAHO

Yellowstone National Park

Yellowstone Lake

Fork

Henry's

Dubois

Ashton

St. Anthony

Teton R.

Mud Lake

Rexburg

Driggs

Rigby

Idaho National Engineering Lab.

So. Fork Snake R.

Arco

Idaho Falls

Jackson

Craters of the Moon National Monument

Palisades Res.

Snake River

Blackfoot

Fort Hall Indian Res.

American Falls Res.

Pocatello

Blackfoot Res.

Portneuf R.

Lake Walcott

Rupert

Soda Springs

IDAHO WYOMING

Burley

Paris

Malad City

Preston

There are few regions of the globe where
the handwriting of nature is larger, plainer,
or less obscured by nature's own
subsequent efforts to erase it.

— *Rossiter W. Raymond*

Report on mineral resources, 1869

SEEING THE LAND

THE SNAKE RIVER begins at 9,200 feet in the icy highlands of the
Continental Divide. Flowing west from Jackson Hole in the Grand Tetons,
it cuts stark, spectacular landscapes. Deep, vertical chasms split barren
plateaus. Wildflowers spread in jagged formations. At Shoshone Falls,
higher than Niagara, the Snake plunges into a gorge, falling 212 feet.

BY TODD SHALLAT

Shoshone Falls, lithograph from the King 40th Parallel Survey, about 1868; previous page, Upper Snake River and Teton Mountains; detail, young Idahoan.

Upstream in eastern Idaho, west and north of Idaho Falls, the river crosses hellish terrain — moonlike craters, dark seas of lava, thick beds of volcanic ash.

Industry, ranching and farming have developed the volcanic basin, but its wilderness traits survive. Explorers called the country "forlorn," "fantastic," "dark and gloomy," "an indescribable chaos," a "frightful glimpse of the Inferno." In 1871 United States geologist Clarence R. King reported

"Forlorn . . . fantastic . . . frightful." An Owyhee canyon.

Ropelike pahoehoe lava with flowering biscuitroot.

desolation. "You ride upon a waste," said King, approaching the "cascading whiteness" of Shoshone Falls. "Suddenly, you stand upon a brink … a great river fights its way through labyrinths of blackened ruins." Today the Snake is a harnessed river. Held by dams, forced through turbines, it generates power for cities and towns in four western states. Tapped for agriculture, it irrigates wheat, barley, sugar beets, alfalfa and nearly one-third of the nation's potatoes. Among the rangeland and farms are dots of population: Boise, Twin Falls, Pocatello, Idaho Falls, all different, all similar — different because each serves a distinct region of a politically factious state; similar because all share water, an arid climate and a history of common struggle with a harsh yet fertile terrain.

Canal surveyors near Milner Dam, about 1907.

To understand the arid crescent is to study the remarkable landscape — its record of human occupation, its flora, fauna and geologic past. Scientists call it the Snake River Plain, a prairie or high desert some 400 miles across. Bounded by Yellowstone to the east and Oregon to the west, it curves below the granite peaks of the Idaho batholith. Geologists see at least two separate regions — a western and an eastern plain. The western section began to appear about 15 million years ago as the Earth broke open in "fissures" that vented molten rock. The eastern section extends from Ha–german to the foot of the Rockies, a bizarre landscape of lava rock and "calderas," or craters, up to 100 miles across. Here geologists study an intense source of heat commonly called a hot spot. According to the most widely accepted theory, the

Balanced Rock near Buhl, 1908.

Shells, about 3 million years old, are delicate evidence of a huge prehistoric lake.

Earth's tectonic plates shifted over this hot spot, melting the crust. The land blistered, spewing out clouds of hot ash. Later, less violent eruptions left tall cones of lava and black fields of rubble along a chain of volcanoes from Craters of the Moon National Monument to American Falls. Today the migrating hot spot causes water to boil to the surface at Yellowstone National Park, evidence of the existence of a giant, potentially active volcano.

Volcanic activity left porous rock that created an unusual drainage system. Snowmelt from the mountains became "lost rivers" that percolated through the surface lava into the Snake River Aquifer, a deep strata of water-bearing rock. As lava flowed into rivers, huge lakes backed up across Idaho and northern Utah. Great floods carved magnificent canyons.

Changes in the natural landscape forced changes in the species that inhabited it. About a million years ago a cooling trend brought Pleistocene mammals to the plain. Elephantine mammoths crashed through the forests. Fierce cats hunted the grassy steppe for elk and bison. There were also warming trends, centuries of hot weather that thawed the tundra and turned the ice into swamp. Fossils near Hagerman preserve the imprint of zebralike horses, small camels, otters, turtles, snakes and a wide variety of shellfish that once inhabited the aquatic plain. Near the close of the Pleistocene Ice Age, perhaps 15,000 years ago, large sloths emerged from marshes to feed in deciduous forests.

Tribes of hunters may have reached the plain about 14,000 or 15,000 years ago. Skilled toolmakers, they were part of the across 16 million square arrivals hunted the last of climate changed, growing

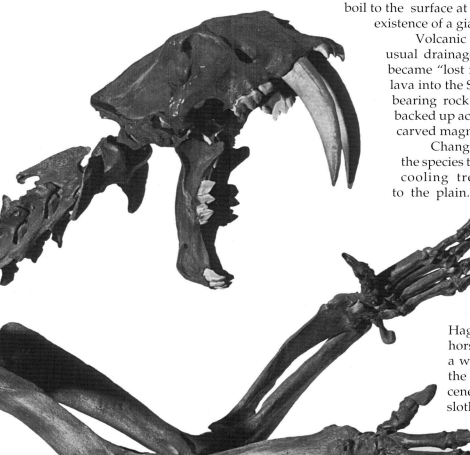

Saber-toothed tiger skeleton

Asian migration that spread humanity miles, from Alaska to Tierra del Fuego. Early the mammoths, driving them into the swamps. As the warmer and more arid, the herd animals thinned, and some species migrated north. Humans diversified. Native Americans, retreating from winter, built shelters in the canyons. In spring they found edible roots just above the plain on the Camas Prairie. Early summer was salmon season along the Boise and Snake rivers.

Combing the foothills in fall, the tribes found seeds and chokecherries to mash into cakes. Spanish horses widened the cyclical migration. By the time of Lewis and Clark, the mounted Shoshone and Bannocks ranged north into Canada and south to the Comanche

lands of the southern plains. Snake River tribes had stone corrals, rock art, villages with grass huts, elaborate hunting tactics and regional economies that moved on and off the plain as it changed according to season.

The reading public first discovered the Snake River country in the words of a man who never saw it himself, the essayist Washington Irving. In *Astoria* (1836), Irving relied on terse observations from Wilson Price Hunt and his party of beaver scouts. Here the basin appeared "cheerless," an empty desert. Irving reconsidered the Snake in *The Adventures of Captain Bonneville* (1837). Vividly descriptive, it told a heroic story of trappers in the wild. The land was now "picturesque," even "romantic." Irving stressed the odd geology of the

"Porcupine Moccasin," a Snake River Shoshone; right, essayist Washington Irving.

Elkskin autobiography of Chief Washakie of the Eastern Shoshone, 1880s. One image (lower left) shows a buffalo hunt. Nine others depict fatal skirmishes with the Blackfeet, Ute and Sioux.

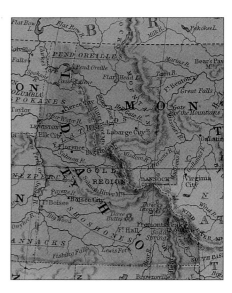

Map Rock (about 1300 B.C.) While probably not a "map" in the modern sense of the word, this petroglyph at Wees Bar in southern Idaho might have been carved to plot or chart constellation changes or herd migrations.

Territorial map (1865). S. Augustus Mitchell's schoolbook geography map reflected a careless attitude toward the West.

Map Rock	Columbus sails	Britain claims Pacific Northwest	Plains Indians acquire horses	Alexander Henry explores Henry's Fork	Frémont maps Snake River
1300 B.C.	1492 A.D.	1579	1700s	1810	1843

Surveyor's transit, about 1936. This 20th-century transit was used to measure changes in elevation for map making and road surveys.

place. The basin was so level it seemed sunken and cracked. Near American Falls and along Bruneau Canyon, the Snake had a dark "volcanic character" with rocks in bizarre formations and cliffs black with basalt.

Soon the volcanic plain was a topic of interest in scientific circles, a frontier for research. The first wave of science was a search for natural wonders. In 1868 geologist King brought a photographer to verify one of the great spectacles of North America: Shoshone Falls. A year later the Smithsonian's Ferdinand V. Hayden found fossils to support his supposition that the plain, once submerged by an ocean, had evolved into freshwater lakes. By the 1890s the focus of science had shifted to precious resources. Geologists soon discovered what prospectors already knew: Snake River gold was abundant, but its flakes were too fine to be easily mined. Water, however, could be tapped through ditches and wells, and in 1902 the U.S. Geological Survey said the potential for hydropower was "practically unlimited." Meanwhile a few wells near Ontario, Oregon, hit small pools of oil and larger deposits of natural gas. In 1920–1921 the frantic search for gushers swept up the Payette and across the plain to the Teton basin. While the Idaho Bureau of Mines and Geology remained unconvinced, calling the plain "unfavorable" for petroleum production, wildcat rigs found just enough natural gas to keep exploration alive.

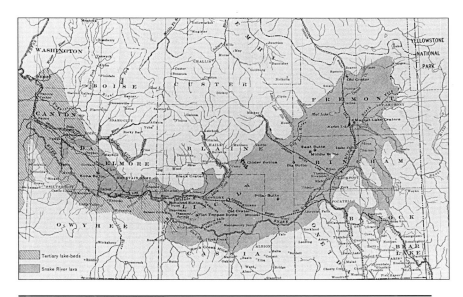

Sketch Map of Southern Idaho (1902). Israel C. Russell, a hydrologist with the U.S. Geological Survey, showed that the Idaho canyon lands were "clothed with vegetation" and included "large tracts of open forest."

Landsat map (1986). Generated by electronic data, this image of Craters of the Moon was produced by a satellite in orbit 600 miles above Earth.

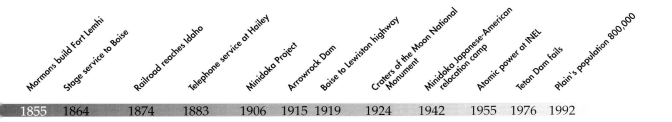

Mormons build Fort Lemhi	Stage service to Boise	Railroad reaches Idaho	Telephone service at Hailey	Minidoka Project	Arrowrock Dam	Boise to Lewiston highway	Craters of the Moon National Monument	Minidoka Japanese-American relocation camp	Atomic power at INEL	Teton Dam fails	Plain's population 800,000
1855	1864	1874	1883	1906	1915	1919	1924	1942	1955	1976	1992

The evolving perspective of science showed that discovery was not an event but a process, an act of the mind as well as the eyes. Explorers, geologists, mining engineers — each group studied Idaho through the distorted lens of its own grand design. The Smithsonian valued the land for its fossils. Oregon-bound emigrants saw a passage through the mountains. Mormons envisioned a poor man's Canaan, a desert Zion. Others found rangelands, veins of coal and phosphate, amazing artesian wells and fertile sites for irrigation.

Some of the grandest promotional schemes have centered on water, Idaho's "liquid gold." Irrigation on the plain began in the 1860s with small diversions from the Boise and Payette rivers. By the 1890s a growing network of canals in the Rexburg-Blackfoot area was already one of the world's most ambitious gravity systems. The 1894 Carey Act granted land for the enormous Twin Falls project at Milner, opened in 1905. Upstream at Minidoka the dam held the U.S. Bureau of Reclamation's first hydro facility, a symbol of the federal commitment to rural electrification. In 1914 the small high school at Rupert was the first in America with electric heat. Meanwhile, reclamation engineers were breaking construction records with the storage reservoir and curved gravity design of Arrowrock Dam on the Boise River. Completed in 1915, the 348-foot dam was the tallest in the world.

Two benefits of federal reclamation were cheap hydropower and the rising water table that created ideal conditions for pump irrigation, a technology pioneered by Idaho irrigators. When Julion Clawson brought drilling equipment to the Minidoka area in 1946, neighbors assumed he was looking for oil. Soon Clawson's electric wells were watering large tracts of beans, wheat and potatoes. Pump irrigation opened the upper Snake River basin to postwar homesteading, and today the plain's groundwater reclaims about 1.6 million acres.

While federal water projects brought wealth to the upper basin, at times the tech–nology failed. On June 5, 1976, the collapse of Teton Dam dramatized the danger of holding water between permeable walls of a lava canyon. In 1984 the discovery of radioactive tritium beneath the Idaho National Engi-neering Laboratory (INEL) forced the lab to shut down a 600-foot injection well. Meanwhile the rising demand for irrigation cut the great river in half — one part impounded by dams above Milner, the rest flowing west and north to the Columbia and the Pacific. In dry years a small child can cross the muddy stream-bed below Milner Dam. Sapped by drought and irrigation, Shoshone Falls at the end of summer stands virtually dry.

The plain's population continues to grow. From 1940 to 1970, the population doubled to 600,000, and today more than 800,000 live in the lava crescent from Ontario to Ashton. Farming and food processing still dominate the desert economy, producing about 75 percent of the state's agricultural returns, but the reliance on irriga-tion leaves Idaho vulnerable to drought and economic slumps. Recessions cut deep in a region so dependent on a few principal crops.

As agribusiness expands, water, Idaho's wealth, remains a perennial problem. Since 1982 a growing dispute between up-stream irrigators and the Idaho Power Company has led to complex adjudication and a challenge to the "first in time, first in line" doctrine of prior

Upper Salmon Falls water wheel under construction, about 1910; Sterling Chalice awarded to the state of Idaho for the best fruit grown under irrigation, National Irrigation Congress, 1903.

*Star Falls, also known as
Caldron Linn, near Murtaugh.*

appropriation, a bedrock of Idaho law. Meanwhile the salmon runs are dangerously close to extinction and Indians have filed lawsuits over treaty fishing rights. Opposition to dams has fueled conservation — a concern for the other uses of water: fish habitats, raft runs, boating facilities, scenic vistas and wildlife preserves. Elsewhere the scrimmage for water has spilled into neighboring states. Since 1964, when Californians began thirsting for a Snake River pipeline, Idahoans have fiercely campaigned to keep the water at home.

Today the contest for water and power mirrors diverse expectations. Some federal agencies still value an empty desert. Isolated, moated by lava, INEL seems a likely site for nuclear research. When the U.S. Air Force looks at the plain near Mountain Home, it sees an excellent training range. Farmers, ranchers and environmentalists resist the Air Force, but their motives are seldom the same. Grazing fees, fire prevention, pesticide use and recreation remain matters of heated debate. One brewing dispute pits ranchers against environmentalists in a conflict over the condition of rangelands. Another conflict centers on the Bruneau River near its confluence with the Snake, where the water table is falling and a small snail was classified as an endangered species before a federal judge overturned that listing on procedural grounds. Meanwhile anglers and dam builders square off over hydro development on the Snake River's Henry's Fork. Each con-

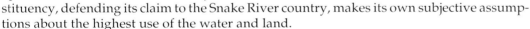

stituency, defending its claim to the Snake River country, makes its own subjective assumptions about the highest use of the water and land.

Thus the Snake country remains a plateau of contrast — blank yet majestic, a dry basin for one of America's largest rivers, a Treasure Valley, a Magic Valley, a shifting mirage. Once a sterile waste, now a fertile crescent — these stereotypes shape our expectations for southern Idaho, yet our images blur as the region evolves. Each generation rediscovers the lava desert. Each revelation of science invites a look at our culture and its ancient debt to the land.

This toothed steppe, furrowed and gouged and spilled in pyramids, is not for persons whose homes are in tropical growth under cloudy skies. This is the last frontier, delivered to rock and desolation and set apart as a monument of its own.

— *Vardis Fisher*

Idaho: A Guide in Word and Picture, 1937

VOLCANIC CRESCENT

CRATERS OF THE MOON, sometime in the future. Steam and volcanic ash rise from the desert. A black flow of slow-moving lava blocks a muddy road. Thin sheets of fluid lava boil up from a 2-mile-long crack in the ground. In places along the crack, fountains of fiery cinders explode in the air. Molten rock, glowing orange, cools into black

BY BILL HACKETT AND
BILL BONNICHSEN

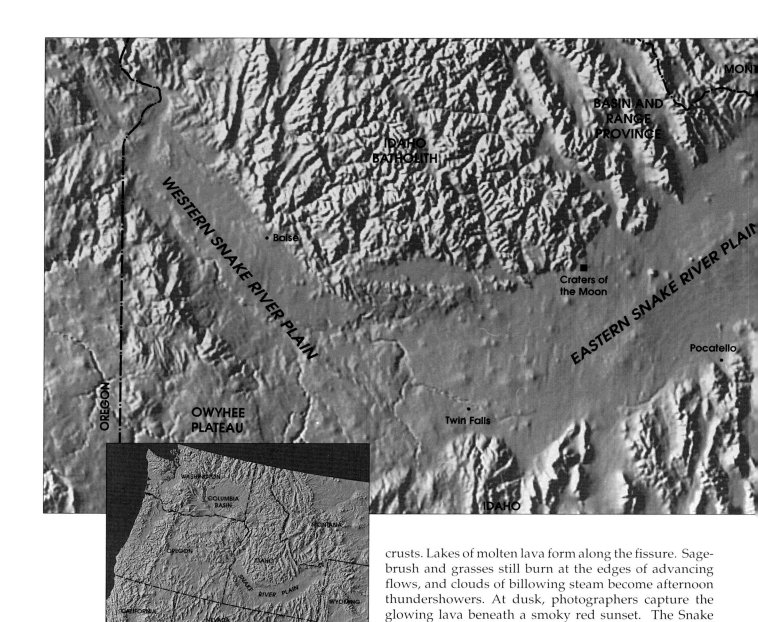

The Snake River Plain
and its relationship to
the Columbia Plateau;
previous page, East
Fork of the Owyhee
River; detail, geologist's
rock hammer.

crusts. Lakes of molten lava form along the fissure. Sagebrush and grasses still burn at the edges of advancing
flows, and clouds of billowing steam become afternoon
thundershowers. At dusk, photographers capture the
glowing lava beneath a smoky red sunset. The Snake
River Plain, dormant for 2,000 years, comes back to life.

The next volcanic eruption — be it five years or 5,000 years from now — will be the latest
episode in a story that has been written over the last 15 million years as the forces of
nature have created the wide crescent of southern Idaho known as the Snake River Plain.

The plain, a 50- to 70-mile-wide belt of sage-covered lava and farmland, is the
dominant geographic feature of southern Idaho. It is also one of the most widely known and
most extensive volcanic regions on Earth. The birthplace of the volcanic belt is near the
shared corners of Idaho, Oregon and Nevada, where the first eruptions occurred about
15 million years ago. Since that time, volcanic eruptions seem to have migrated eastward at
1 to 2 inches per year, with the most recent volcanic activity having taken place in eastern
Idaho and Yellowstone. Viewed in terms of geologic time and the age of the Earth, the
formation of the Snake River Plain–Yellowstone region is a relatively recent event, repre-
senting only a few tenths of one percent of the Earth's history.

Eruptions on the plain have come in cycles. For example, the Craters of the Moon lava
field was formed by dozens of lava flows during eight eruptive periods over the past 15,000

Yellowstone Plateau

Island Park

Teton Range

WYOMING

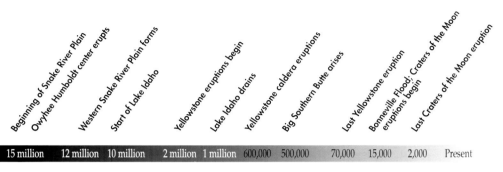

Shaded relief map of Snake River Plain produced on computer using digital topography. From U.S. Geological Survey map I-2206.

Beginning of Snake River Plain

Owyhee Humboldt center erupts

Western Snake River Plain forms

Start of Lake Idaho

Yellowstone eruptions begin

Lake Idaho drains

Yellowstone caldera eruptions

Big Southern Butte arises

Last Yellowstone eruption

Bonneville Flood; Craters of the Moon eruptions begin

Last Craters of the Moon eruption

| 15 million | 12 million | 10 million | 2 million | 1 million | 600,000 | 500,000 | 70,000 | 15,000 | 2,000 | Present |

years. Each eruptive period lasted only a few years or decades and was followed by a long, quiet interval of a thousand years or more. These accumulated lava flows have buried the Snake River Plain to depths ranging from several hundred feet thick in southwestern Idaho to more than a mile thick on the eastern portion.

But what have the lava flows buried? Beneath the thousands of feet of blue-black basalt are the remnants of massive volcanoes — just like the caldera that occupies Yellowstone Park. The basalt lava flows we see on the surface are a relatively thin veneer covering these ancient volcanoes and the enormous deposits of tuff they created.

Most geologists believe the Snake River Plain volcanic belt was caused by the westward movement of the North American continent over a source of magma, or "hot spot." These hot spots are fed by deep, upwelling currents of extremely hot, perhaps partially molten rock. Creeping upward at rates less than 1 inch per year, these volcanic plumbing systems, known as mantle plumes, can extend several hundred miles into the Earth's interior and are the source of basaltic magma that erupts at hot-spot volcanoes.

Deeply rooted in the Earth's mantle, the plumes essentially remain fixed in place for tens of millions of years while the continental plates shift above them. Like burn marks that form on a piece of paper as it is moved over a candle flame, hot-spot volcanoes erupt onto the Earth's surface above these mantle plumes. Above any one hot spot, volcanism is active for a few million years and the volcanoes are slowly carried away from the heat source as the Earth's crust shifts. Eventually they become extinct and are covered by subsequent lava flows and sediments.

Mantle plumes leave trails of hot-spot volcanoes that become progres-sively older as they move away from the plume. Thus, as the North American continent moved westward, the trail of volcanoes on the Snake River Plain that resulted from the mantle plumes moved eastward. The hot spot is believed to

BASALT, the most common type of lava, is high in iron and low in silicon and volatile gases. It typically comes to the Earth's surface in mild eruptions, with the consistency of honey or peanut butter.

ASH AND TUFF FLOWS are created when rhyolite magma under tremendous pressure erupts in great explosions which send fountains of molten ash into the sky. As the ash falls back to Earth it spreads laterally and compacts under heat and pressure, welding together in layers of tuff.

The **MANTLE** is a zone of the Earth beneath the crust, starting 30–40 miles beneath the surface.

MAGMA is molten rock resulting from the melting of the upper mantle or deep crust.

Once molten rock erupts onto the Earth's surface, it is called **LAVA**.

At the Idaho-Oregon border, where the Snake River Plain eruptions began 15 million years ago, superheated clouds of ash erupted in great explosions and cascaded across the terrain. When the flows stopped, the ash fused, creating deep layers of welded tuff.

have originated at what is now the junction of Idaho, Nevada and Oregon. The migration of the continent over the hot spot created a line of huge explosive volcanoes trending northeast. Geologists view southwestern Idaho as the oldest expression of the hot spot, and Yellowstone Park as the most recent.

The early volcanic activity was much more violently explosive than the relatively calm basalt lava flows that followed. The explosiveness of this early volcanic period is attributed to the type of rock being melted and put under pressure. The hot spot melted the Earth's lower crust, creating rhyolite, a type of magma containing dissolved water vapor and much silica. Rhyolite eruptions can be very explosive, with ash discharge 100 to 1,000 times more voluminous than the 1980 eruption of Mount St. Helens.

Frequently on the Snake River Plain, the volcanoes erupted in great explosive clouds of hot, gas-charged rhyolite ash that shot 5–10 miles into the sky. The superheated ash returned to Earth and flowed for miles, then bonded together, forming what is known as welded tuff. Once the explosive rhyolite magma had erupted, the hotter and heavier basalt magma rose to the surface, burying the rhyolite under successive flows that spanned several million years.

Layers of rhyolite pumice and ash flows can be seen in the walls of the Snake River canyon. Overlying these layers, exposed in cross section by the river, are layers of black basalt, which mark the second wave of volcanic activity on the plain.

In the wake of the hot spot, the Earth's crust has deflated into a long trough that is filling with sediments eroded from the surrounding mountains. Thus, the Snake River Plain can be thought of as a burial ground of ancient Yellow–stone "national parks," each taking its place in line, awaiting coverage by younger lava and sediments.

Rhyolite flows on the Owyhee River canyon exhibit cooling fractures which, when eroded, yield talus slopes of large red rhyolite plates.

RHYOLITE magma is created by the melting of the continental crust at great depths. Rhyolite is thick magma high in silicon and volatile gases, which cause it to erupt violently as it reaches the Earth's surface.

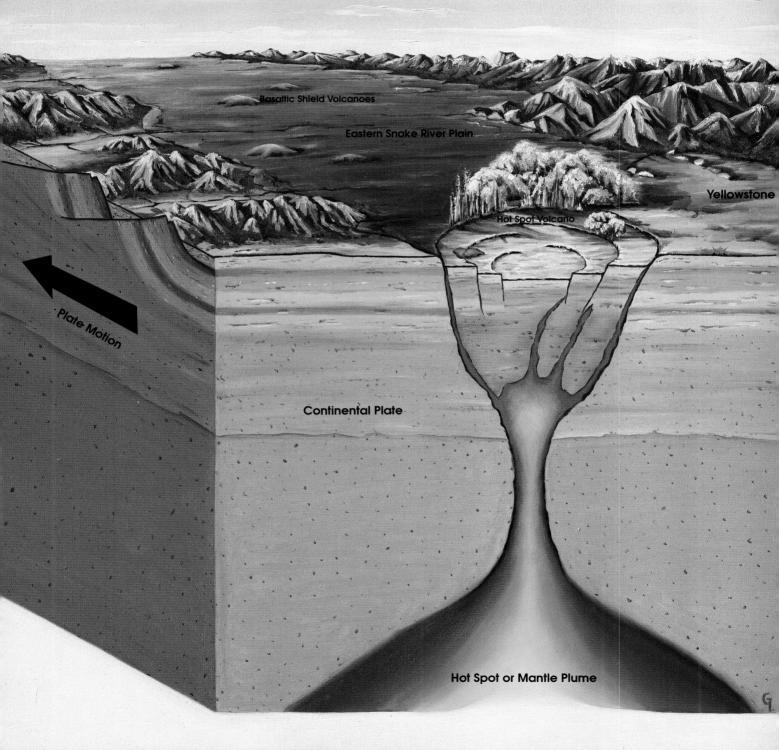

Yellowstone National Park is the latest expression of the "hot spot" that is believed to have created the Snake River Plain. The hot spot is a stationary plume of partly molten rock from deep within the Earth. As the westward-moving continental plate passes over the plume, the continental crust is melted and magma erupts to the surface in tremendous explosions. The resulting calderas — huge craters — are later buried by sediments and flows of basalt magma.

Basaltic Shield Volcanoes

Eastern Snake River Plain

Yellowstone

Hot Spot Volcano

Plate Motion

Continental Plate

Hot Spot or Mantle Plume

VOLCANIC PROCESSES: A PRIMER

Few natural events are more awesome than volcanic eruptions, which give us brief glimpses into the internal workings of the planet. The most ancient rocks on Earth are of volcanic origin; the crust of the Earth also is largely volcanic rock. Together with life-giving oxygen from photosynthetic plants, volcanic gases have produced the Earth's atmosphere ... and continue to add to it. Although most people think of volcanoes as symmetrical, snow-capped cones with ash billowing from their summits, these are only one variation.

Some volcanoes hardly look like mountains, but all have an essential ingredient — molten rock called magma. Most magma originates from the melting of the Earth's deep crust and upper mantle. Upon reaching the surface, the magma, generally known as lava, takes many forms depending on its composition and properties.

Basalt lava, the most common type on Earth, is dark due to its high iron content. The countless basalt volcanoes that erupt unseen beneath the ocean add about 1 cubic mile of new lava each year to the Earth's crust. Blocked by the thick continental crust, basalt eruptions are less common on land, but they do occur where continents are being torn apart during the creation of new ocean basins. Basalt also erupts in places like the Snake River Plain, where hot plumes of magma melt through the crust. Basalt typically reaches the surface as mild effusions of thin, fluid lava flows.

The Hawaiian word "pahoehoe" is used for the smooth, ropy-surfaced lava. Pillow lava is another basalt variety that forms underwater, where it develops a glassy skin. From 6 inches to 6 feet across, each pillow of lava resembles piles of extruded toothpaste or deflated beachballs. The spaces between the pillows are usually filled with pieces of a pea-sized glassy material.

Rhyolite, another volcanic rock, is a lava that is formed during the melting of the deep continental crust. It is typically pale and sugary; less commonly, it is glassy, in which case it forms black obsidian.

As opposed to basalt, rhyolite moves sluggishly and usually solidifies underground, forming huge bodies of crystalline granite that break the Earth's crust and slowly cool to form so-called plutons and batholiths. When rhyolite does erupt, however, expanding gases can literally blow the roof off the magma chamber, and entire continents can be blanketed with ash from the ensuing explosions.

LAVA TYPES

As the underground rhyolite magma chambers are evacuated during eruptions, the overlying roof rocks often collapse inward like a fallen soufflé, forming a broad, shallow depression up to tens of miles across and up to thousands of feet deep. Known as calderas, these giant, circular depressions may have spectacular, steep walls such as Crater Lake in Oregon. Others are so broad and shallow that they are hardly recognizable as volcanoes. Filled by ash and pumice, they are nonetheless the sites of volcanism on its grandest scale.

In the long intervals between explosive eruptions, relatively quiet rhyolite eruptions commonly occur. Viscous lava flows ooze from fractures in the Earth's crust. The pasty lava crystallizes very sluggishly and is often chilled to form obsidian, a dense variety of volcanic glass.

Huge, explosive eruptions of rhyolite are relatively infrequent in the geological record. On the Snake River Plain, they have occurred about once every million years or so. None has occurred during historic times.

Andesite lava, a third variety, is a fine-grained derivative of basaltic magma that forms small cinder cones and thick block-lava flows in places like Cedar Butte and Craters of the Moon. The beautiful volcanoes along the Pacific, the "Ring of Fire," are andesite cones, but andesite is rare on the plain.

A'A is the rough lava created as flowing lava cools and becomes viscous.

Pahoehoe is a smooth, ropy lava created when the surface of a basalt flow cools and the fluid material beneath keeps moving.

Blue dragon is pahoehoe lava containing titanium oxide, which creates an irridescent blue sheen.

Pillow lavas are formed when basalt flows into water. The outer surface cools into a glassy skin.

Andesite is intermediate in composition between basalt and rhyolite. Its pasty consistency forms rubbly surfaces and spines.

Obsidian is volcanic glass created by rapidly cooling, gas-free rhyolite lava.

VOLCANIC LANDFORMS

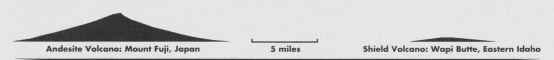

Andesite Volcano: Mount Fuji, Japan   5 miles   Shield Volcano: Wapi Butte, Eastern Idaho

Rhyolite Volcano: Bruneau-Jarbidge Eruptive Center, Southwestern Idaho

The violence of volcanic eruptions depends on two components in magma: silicon and dissolved gases. The silicon content controls the viscosity of magma — its pastiness or resistance to flow. As with adding flour to pancake batter, increased amounts of silicon in the magma serve to increase viscosity. Water vapor is the major gas dissolved in most magma. Rising magma can also boil groundwater and touch off explosions that rip through the crust.

In principle, these eruptions resemble the opening of a bottle of soda pop, where high-temperature steam takes the place of the carbonation bubbles and viscous rhyolite magma the place of the soda pop. In the sealed soda pop bottle, gases are held invisibly in solution, which is analogous to water vapor dissolved in magma deep underground (such magma never contains gas cavities). When the bottle is opened, the expanding gas bubbles usually float gently to the surface, as they do in quiet volcanic eruptions of fluid, gas-poor magmas. But if the liquid is

under great pressure and contains an excess of gas, as when the sealed contents are shaken, a liquid-gas froth may gush from the bottle. In a volcanic explosion, the internal pressure within the viscous, gas-charged magma is so great that the gas expands explosively, tearing the magma into bits of glassy froth called pumice, or bursting the bubbles themselves, resulting in tiny shards called volcanic ash.

That ash may spread for miles, settling and welding together to create deposits of tuff hundreds of feet deep. Usually these tuff deposits are buried by subsequent, less violent flows of rhyolite or basalt.

But where rivers have carved deep canyons, these layers of different lava flows can be seen in sequence, descending back into time as one views the walls from top to bottom. Viewed this way, our farms and towns perched atop the latest flows of basalt can be seen for what they are: a tiny episode of occupation on a land that has continually re-created itself by burying its former surfaces.

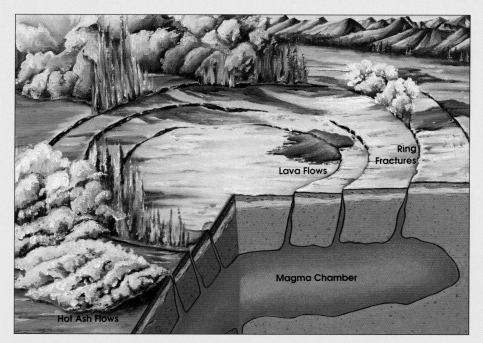

Caldera formation. Rhyolite magma rising from the depths bulges through the Earth's surface, then breaks through in a cataclysmic eruption which sends molten ash thousands of feet into the sky. Following the eruption, the now-fractured dome collapses into the evacuated magma chamber, forming a huge crater or caldera that may be tens of miles across and several thousand feet deep.

VOLCANIC ORIGINS

The Owyhee-Humboldt eruptive center, the area in which the first major eruptions of rhyolite occurred on the Snake River Plain, does not look like a typical volcano. Rather, it is a broad, dishlike area of about 40 by 60 miles that apparently collapsed during rhyolite eruptions. An obvious caldera did not form, perhaps because the magma chamber was too deep to allow the overlying rocks to cave in. At the central part of the Owyhee-Humboldt eruptive center is a field of basalt shield volcanoes and accompanying lava flows that are about 10 million years old. These are surrounded on all sides by older layers of rhyolite — mainly ash-flow tuffs — that are tilted inward toward the center of the basin, beneath the basalt.

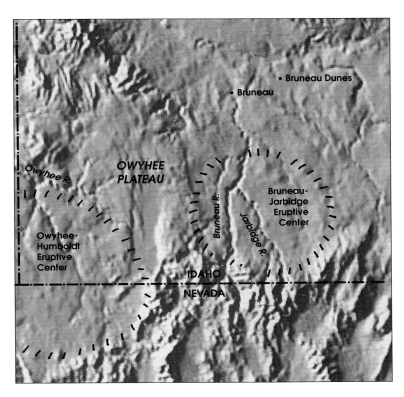

About 11 million years ago, volcanism shifted to the east, forming the Bruneau-Jarbidge eruptive center. Like the older Owyhee-Humboldt area, the Bruneau-Jarbidge center is a broad, downwarped region some 60 by 35 miles across. Between about 10 to 11 million years ago, a sequence of 11 or more welded, ash-flow tuff layers (the Cougar Point Tuff), at least 12 large rhyolite lava flows, and many thin basalt lava flows erupted within the basin. The eastern and northern margins of the eruptive center are now buried beneath the younger basalt lava flows of the central Snake River Plain. Farther south, the Cougar Point Tuff is magnificently exposed in the deep canyons of the Bruneau and Jarbidge rivers.

Each ash-flow deposit of the Cougar Point Tuff was typically formed by several glowing clouds of ash and pumice, spaced so closely in time that they welded and cooled as a single layer of tuff. Near the base, particles of volcanic ash were densely welded into obsidian or vitrophyre (obsidian with crystals). The emplacement of each layer was a distinct volcanic event, and buried soils and sediments between layers of welded tuff indicate the passage of considerable time between these cataclysmic eruptions.

About a dozen large rhyolite lava flows are located within the Bruneau-Jarbidge eruptive center, and several of them are exposed in deep canyons there. One of the most-studied flows is the

The Jarbidge River Canyon exposes layers of welded tuff and rhyolite flows. These flows are from the earliest rhyolite volcanic eruptions of the Snake River Plain.

The Owyhee River, like the Jarbidge and Bruneau rivers, was carved when Lake Idaho was suddenly drained by Hells Canyon of the Snake River. The canyon had eroded headward, creating an outlet for the huge body of water. The Owyhee River, which drained into Lake Idaho, then began eroding headward as its gradient was suddenly steepened by the dropping water level of the lake.

The Cougar Point Tuff, a sequence of strongly welded ash-flow layers in Bruneau Canyon, is eroded into pillars. In the distance, the upper canyon wall is created by a 300-foot layer of the Cougar Point Tuff.

Dorsey Creek Rhyolite, a single lava flow that is exposed in the walls of the Jarbidge River canyon for a distance of about 25 miles. This flow has a minimum volume of about 20 cubic miles and is at least 700 feet thick.

Large-volume rhyolite lava flows similar to those of southwestern Idaho and Yellowstone are known in few other regions of the world. Because of the extensive exposures in the deep canyons of the Bruneau and Jarbidge rivers, the shapes and internal features of these unusually large lava flows have been well documented.

Rhyolite lava-flow interiors generally have three zones: a quickly cooled basal zone of glassy or chaotically broken rhyolite; a thick central zone of gray to reddish brown, finely crystalline rhyolite that formed during relatively slow cooling; and an upper cap and margins of sheeted, contorted and broken rhyolite. The large quantities of broken rubble on the tops and margins of rhyolite lava flows are formed as a chilled, brittle skin develops on the moving lava. The massive interiors of flows are typically more than 150 feet thick.

More than half of the Bruneau-Jarbidge eruptive center is covered by basalt lava flows, each ranging from 5 to 30 feet in thickness. The basalt lavas erupted from shield volcanoes scattered around the area. About 40 shields are known, each made of thin basalt lava flows and with a typically low profile. Most are circular or oval, with heights from 50 to 300 feet and diameters typically between one and four miles. Many have craters preserved at their tops. Since buried soils and sediments do not occur between the lava flows that erupted from a single volcano, it appears that the succession of basalt lava flows from each volcano erupted in only a few years. Much of the basalt evidently spread as thin sheets as a consequence of rapid discharge of fluid lava over relatively flat terrain.

CLARENCE R. KING
(1842–1901)

He was "the best and brightest man of his generation," said historian Henry Adams; "the richest and most many-sided genius of his day." Clarence King of Yale — poet, mining engineer, a founder and the first director of the U.S. Geological Survey — was a hero from a golden age when the Intermountain West was the frontier of American science. An urbane man with a derby and bright yellow gloves, King, at age 25, organized the famous 40th Parallel Survey, a study of the high desert between the Rockies and the Sierra Nevada. The expedition reached Idaho in 1868. "The monotony of the sage desert was overpowering," King reported, yet the beryl-green Snake with its ragged cliffs was "dazzling" and "fantastic," a scene of "extreme beauty." King's investigation of volcanism in Idaho confirmed his belief that the Earth, younger than previously thought, was no older than 24 million years. His Idaho studies also contributed to a classic work on the western mountains, *Systematic Geology* (1878), a sweeping overview that showed how quakes, floods and eruptions altered the biological process of natural selection and accelerated the pace of ecological change.

THE WESTERN PLAIN

The western Snake River Plain, a broad valley with Boise at its northeastern margin, is perhaps best considered as a troughlike structure that developed at the same time as other geologically young mountains of southern Idaho and northern Nevada. It began forming about 12 million years ago with major eruptions of rhyolite, followed by voluminous outpourings of basalt lava.

Large amounts of basalt lava, together with lake and stream sediments, were deposited onto the western plain as it slowly subsided during stretching of the Earth's crust. In the lower reaches of Bruneau Canyon, some 800 feet of basalt lava flows are exposed. Farther to the northwest, lake and stream sediments dominate the surface geology, but deep drilling has revealed that much basalt lava is buried far below the Earth's surface, indicating that several thousand feet of subsidence, or sinking, has occurred on the western Snake River Plain during the last few million years.

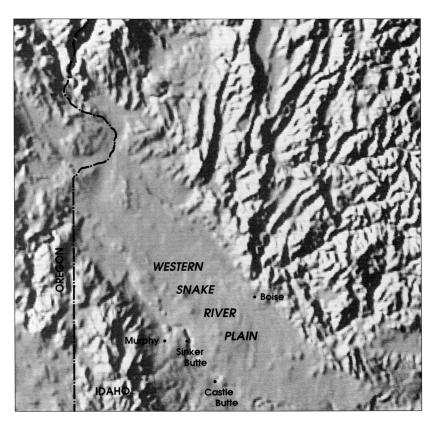

Considerable geologic evidence, including lake-deposited sediments, fossils and pillowed basalt flows, indicates that large lakes blanketed southern Idaho and filled the troughlike basin of the western Snake River Plain between 1 and 10 million years ago. Lake Idaho is a geological term for this series of lakes that waxed and waned with the changing climate.

In the lower Bruneau Canyon, as well as in other areas throughout the western and central Snake River Plain, features of the basalt lava flows and pyroclastic deposits suggest that the highest stand of Lake Idaho was about 3,800 feet above sea level. Lake Idaho was permanently drained 1 million years ago when a then-small tributary of the

PYROCLASTICS are fragments of magma caused by expanding gases trapped in the lava. Basalt pyroclasts frequently fall to the ground in piles, building cinder cones.

Lake Idaho built a reef of algal mounds and snail shells, which is exposed near Bruneau in a rock formation known as the Hot Spring Limestone.

EDWARD D. COPE (1840–1897), a leading vertebrate paleontologist and a president of the American Association for the Advancement of Science, found Pliocene fossils of mollusks, crayfish and other marine life that once inhabited Idaho's marshes and lakes.

Salmon River eroded headward, eventually reaching the lake and causing its waters to drain northward, forming the Snake. This tributary, eroded by the massive flood of water from Lake Idaho, became the Hells Canyon of the Snake River.

During and after the final lake recession, the Bruneau River and other streams that emptied into the lake were left with mouths perched high above the lake floor. As these streams eroded downward, adjusting to the new base level, they carved deep canyons through the basalt. Thus the many magnificent canyons in southwestern Idaho, including those of the Owyhee and Bruneau rivers, owe their origins to the demise of enormous Lake Idaho.

During the era of Lake Idaho, volcanic eruptions and lava flows continued, with the interactions of magma and water creating unique geological features that can be seen throughout southwestern Idaho.

Explosive interaction of basalt magma and underground water and lake water produced large cones of fine-grained tuffs. The distinctive orange-brown color of these basaltic tuffs is due to palagonite, a substance that forms when hot basaltic glass chemically interacts with steam or water. Deposits of palagonite tuff can reach hundreds of feet thick near large volcanoes such as Sinker Butte. Some of these tuff cones built up above the lake level, and their eruptions changed from explosive magma-water reactions to Hawaiian-style fire fountains. Often the explosions were powerful enough to throw yard-wide blocks and bombs of basalt hundreds of feet from their source. The "caps" of welded spatter on Sinker Butte and Castle Butte protected the softer tuffs from erosion, so they remain as prominent landforms today.

Where the lava poured down underwater slopes, it formed bulbous masses of glassy-skinned rock — pillow lava. Thick accumulations of pillow lava made extensive deltas along some of the steeper shorelines. Other lava flows were altered into crumbly brown masses that look as if the rock had "stewed" in warm water.

WELDED SPATTER is an accumulation of lava bombs and ejected magma which melts together and cools, forming pillars and cones or capping previous volcanic eruptions.

This cross section of a volcanic crater was created when Sinker Creek eroded through the crater margin of the Montini volcano near Murphy. Basaltic tuffs were erupted from a large crater to the left and piled several hundred feet high. The cap of welded spatter on the top of the hill has protected the softer tuffs below from erosion.

THE EASTERN PLAIN

"The route was very rocky," said trapper Osborne Russell, crossing the plain in 1835:

On surveying the place I found I could go no further in a south or east direction, as there lay before me a range of broken, basaltic rock which appeared to extend for five or six miles on either hand and five or six miles wide, thrown together promiscuously in such a manner that it was impossible for a horse to cross them. I had plenty of provisions, but could not eat. Water! Water was the object of my wishes. Traveling for two days in the hot burning sun without water is by no means a pleasant way of passing time. I soon fell asleep and dreamed again of bathing in the cool rivulets issuing from the snow-topped mountains.

Russell and his contemporaries would have been surprised to learn that the barren land of the eastern plain would one day be irrigated by underground water pumped from buried basalt lava flows and sedimentary deposits.

The eastern Snake River Plain is underlain in places by more than a mile of basaltic lava flows, erupted mostly within the last 2 million years. Within the fractures and cavities of the basalt lava flow courses one of the eastern plain's most remarkable features — the Snake River Aquifer.

The aquifer is a huge network of underground water underlying about 10,000 square miles of south central and eastern Idaho, from the town of Hagerman northeast to Island Park. The aquifer is fed by rainfall and melting snow from a 35,000-square-mile area in the surrounding basin and range mountains. Several of Idaho's "lost rivers" meander for miles over permeable lava beds before they percolate into the ground. The eastern and central Snake River has only two northerly tributaries — Henry's Fork of the Snake and the Big Wood River. All other northerly streams disappear into the lava and feed the aquifer before reaching the Snake River. The underground water moves slowly through the volcanic rocks and sediments, at rates estimated to be 2 to 10 feet per day. Yet the yearly influx and discharge of underground water is substantial: about 8 million acre-feet — enough water to cover the entire state of Idaho in 1.5 inches of water. The total storage capacity may be equal to the present volume of Lake Erie, enough water to exceed four centuries of Snake River flow.

The aquifer feeds the Snake River in many places, notably at springs near American Falls and between Twin Falls and Hagerman. The Thousand Springs area near Hagerman contains 11 of the 65 largest cold springs in the United States. Between 1910 and 1950 the volume from the springs almost doubled due to upstream irrigation — ditch water drawn from the river that percolated into the aquifer

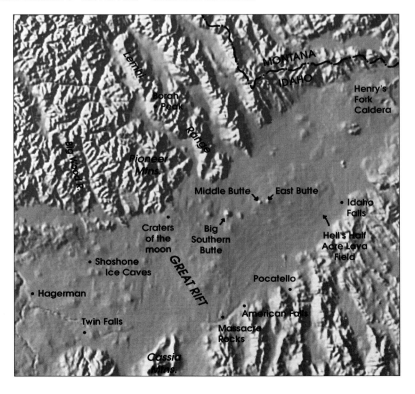

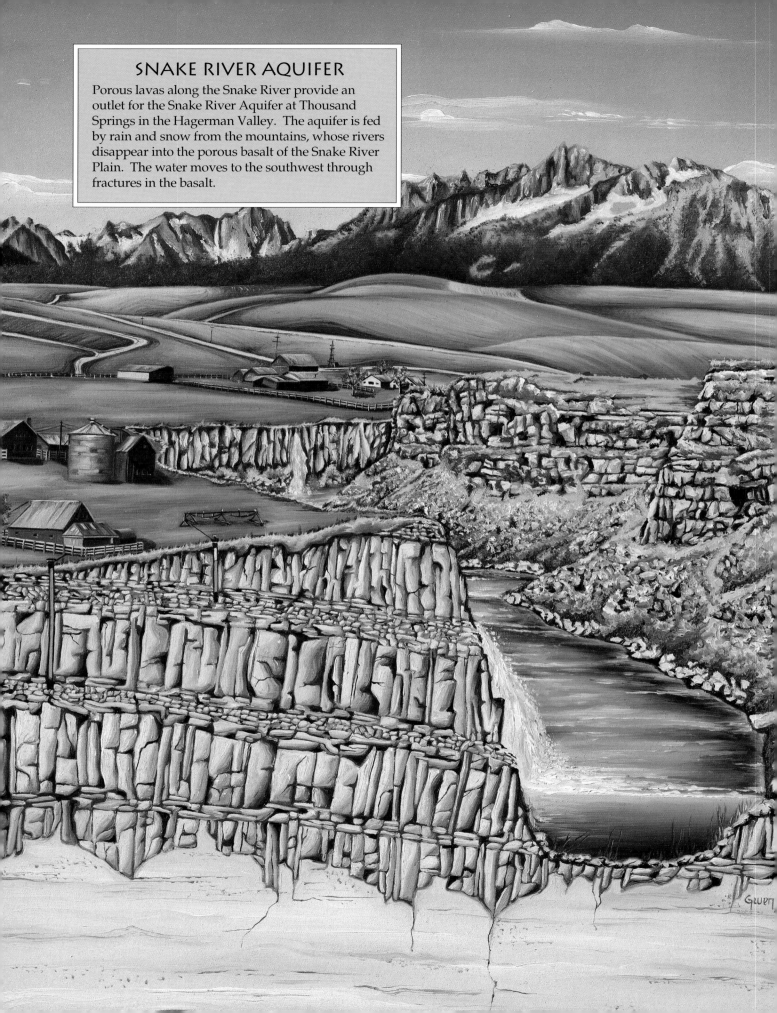

SNAKE RIVER AQUIFER

Porous lavas along the Snake River provide an outlet for the Snake River Aquifer at Thousand Springs in the Hagerman Valley. The aquifer is fed by rain and snow from the mountains, whose rivers disappear into the porous basalt of the Snake River Plain. The water moves to the southwest through fractures in the basalt.

below. The discharge has since declined, due mainly to the increased use of groundwater for irrigation and to decreased diversions of surface water.

From April to October each year, nearly all of the Snake River water upstream from Twin Falls is diverted for irrigation, and aquifer-fed springs provide most of the river water that is used downstream for agriculture, aquaculture, power generation and residential consumption. A few impressive springs remain, although commercial development has diminished the former grandeur of the area.

The volcanic and hydrologic processes that created the Thousand Springs area have recurred many times on the eastern plain. North and east of what is now the town of Hagerman, basalt volcanoes were periodically active during the past half million years or so, filling several ancestral Snake River canyons with lava and forming temporary lakes. Later lavas that flowed into the lakes were highly fractured and created permeable pillow lavas, whereas those in the dry channel downstream of the dams were relatively dense and impermeable.

The Snake River has since eroded its present canyon farther to the south, cutting through the permeable pillow lavas and unleashing the groundwater contained in them. The emergence of springs is thus controlled mainly by the distribution of pillow lavas and other, less permeable lavas and sediments in the north canyon wall.

Geologic evidence of the river's shift has been reported from many places on the plain. Pillow lavas and lake beds are exposed in the canyon of the Snake River near Bliss and Hagerman and along the Boise River canyon. On the eastern Snake River Plain, American Falls Lake was an ice-age lake that formed behind lava flows from vents near Massacre Rocks. Many well-preserved fossils of large ice-age animals have come from the ancient sediments of American Falls Lake.

The Thousand Springs area of the Hagerman Valley includes 11 of the 65 largest cold springs in the United States. All but a few have been developed for aquaculture and hydropower.

Basaltic volcanism was generally very mild in character, with lava flows erupting from small shield volcanoes and fissures. Sedimentary deposits interlayered with the lava flows include sand dunes, wind-blown silt (known as loess), playa deposits, lake beds and river sediments. The eastern plain is dotted by dozens of shield volcanoes, each covering a few square miles and built by thousands of thin, fluid lava flows. Cinder cones

are relatively uncommon on the eastern plain, and most are confined to volcanic rift zones where viscous, gas-rich magma erupted, such as at the Craters of the Moon lava field.

Many flows were fed by extensive systems of lava tubes — underground channels within the flows that developed when the lava crust congealed and insulated the still-fluid interior of the flow. When lava ceased to flow from the vent, many of the tubes drained and their roofs collapsed, leaving an underground network of tunnels.

Basalt lava flows on the eastern plain cover an even thicker and much larger accumulation of older rhyolite deposits, exposed around the edges of the plain, from the Cassia Mountains to the Teton Range on the southern margin, and from the Centennial Mountains to the Pioneer Mountains along its northern margin.

One place where the cover of basalt is sufficiently thin for the Snake River canyon to cut deeply enough to expose the underlying rhyolitic volcanic rocks is 212-foot-high Shoshone Falls, about 4 miles from the city of Twin Falls.

Shoshone Falls was probably created during the Bonneville Flood, about 15,000 years ago, when the natural dam holding back the waters of Lake Bonneville gave way, unleashing a monumental torrent which filled the river canyon. During the flood, the Snake River eroded

Melon boulders of basalt were ripped from the walls of the Snake River canyon during the Bonneville Flood, rounded by bumping against one another and deposited in gigantic "gravel" bars several hundred feet above the Snake River.

The 212-foot-high Shoshone Falls — higher than Niagara — was created during the Bonneville Flood as the river eroded during the cataclysmic deluge.

The historic photograph above shows the awesome spectacle of Shoshone Falls. But in recent years the Snake River has been literally de-watered upstream for irrigation, leaving the falls with but a trickle of water, as seen in this aerial view below.

*Lava river
on the Great Rift*

headward, cutting the canyon deeper upstream. The flood was as short-lived as it was immense. When the lake was drained and the flood subsided, the erosion halted. Shoshone and Twin Falls mark the headward erosion of the canyon when the flood ended.

Deeply eroded volcanic rocks in the area around the falls are evidence of a great deluge. The floor of the canyon even today is still strewn with boulders from the flood.

THE HEISE VOLCANIC FIELD

On the eastern plain, geologists have identified three very large and extensive rhyolite ash-flow sheets, ranging in age from 4.3 to 6.6 million years. Together with other volcanic rocks and sediments, those three major ash flows form the Heise volcanic field.

Rhyolitic deposits of the Heise volcanic field cover more than 13,500 square miles in southeastern Idaho. The three widespread rhyolite ash flows include the 6.6 million-year-old Blacktail Tuff, the 6 million-year-old Walcott Tuff and the 4.3 million-year-old Kilgore Tuff. Volcanic ash layers from these eruptions are found across much of the western United States.

All three of these ash flows erupted from major calderas that are now largely buried beneath younger basalts and sediments formed mostly during the last 2 million years.

*The 4.3 million-year-old
Kilgore Tuff seen in the
foreground is one of three
widespread ash flows from
the Heise volcanic field on
the eastern Snake River
Plain. The Lemhi Range can
be seen in the background.*

THE GREAT RIFT

In places on the eastern Snake River Plain, basalt volcanoes form lines of fissure-fed lava flows, small shields, spatter and cinder cones, pit craters and open cracks called volcanic rift zones.

One of the best known is the Great Rift, a 60-mile-long zone that crosses the eastern plain from the Snake River near American Falls to Craters of the Moon National Monument. Most of the outpourings were of basaltic lava, but some of the pyroclastic cones at Craters of the Moon were produced by more violent eruptions of viscous andesite lava. Carbon-14 dating of burnt vegetation beneath the lava flows and paleomagnetic dating have shown that volcanism along the Great Rift occurred mainly within the past 15,000 years, and the youngest eruptions occurred only about 2,100 years ago. Three lava fields were formed, separated from each other by noneruptive or "dry" rift segments, where the ground was cracked open by underground magma, but no eruptions took place.

Along a 6-mile segment of the northern Great Rift, the Craters of the Moon lava field was formed by dozens of lava flows and pyroclastic cones from eight eruptive periods that occurred over the past 15,000 years. Each eruptive period probably lasted a few decades or centuries, and each was followed by a long, quiet interval of a thousand years or more. Lava and pyroclastics totaling about 8 cubic miles were erupted from within the Craters of the Moon area, one of the largest postglacial lava fields in the world.

South of Craters of the Moon, the Great Rift is marked by a 25-mile-long segment of open cracks from which no lava issued during the last 15,000 years. A small field of thin basalt lava flows developed at the south end of the open cracks about 2,200 years ago. Known as the Kings Bowl lava field, this was the site of a lava pond that was fed by a 4-mile-long eruptive fissure. When the eruptions ended, lava drained out of the fissure, allowing groundwater to flood into the crack. The resulting steam explosions formed a 30-yard-wide explosion crater, known as Kings Bowl. Drainage of lava has left a complex system of underground caverns deep in the Kings Bowl rift segment. Known as the Crystal Ice Caves, they are a natural refrigerator, trapping cold, dense winter air that maintains the caves at a temperature just below freezing. Impressive pools and columns of ice occur in multiple chambers along the rift, fed by 2 millennia of rain and snow that have dripped down into the frozen caverns.

The southernmost volcano along the Great Rift is Wapi Butte, a basaltic shield volcano that last erupted about 2,300 years ago, and produced nearly 1.5 cubic miles of thin, fluid lava flows.

HAROLD T. STEARNS
(1900–1986)

The Hawaiian Islands look nothing like the arid Snake River country, but USGS geologist Harold Stearns, an authority on Idaho and the South Pacific, explored the connection: both were volcanic regions where water flowed freely underground through strata of fractured basalt. Stearns' research on Hawaii began with the great eruption of Kilauea Volcano in 1924. His Snake River work ranged from oil and diamond exploration to the Hagerman fossil beds. Stearns also won high praise from President Calvin Coolidge for the first scientific investigation of the Craters of the Moon area. From 1947 to 1959 he mapped irrigation and hydroelectric projects downstream of Pocatello — 11 dam sites in all.

Great Rift: left, cinder cones mark the line of the volcanic rift zone in the Craters of the Moon National Monument; above, Highway 20 cuts across a tongue of lava near Craters of the Moon; lower left, an eruptive fissure along the Great Rift; below, a cinder cone at Craters of the Moon formed during the most recent eruptions on the plain 2,100 years ago.

EXPLOSIVE BASALT VOLCANISM

Although basalt volcanoes normally erupt mildly, explosive basalt volcanism has occurred on the eastern plain due to the interplay between rising basaltic magma and underground water. Pyroclastic cones and explosion craters occur in the Massacre Rocks area near American Falls, the Menan Buttes near Rexburg, the Kings Bowl crater along the Great Rift and several other scattered vents.

The Menan Buttes are basaltic tuff cones that erupted along a 3-mile-long fissure about 8 miles west of Rexburg. The cones were formed during subsurface injection of basaltic magma into water-saturated river gravels and basalt lava flows of the aquifer. The resulting explosions threw out sand-sized brown tuff, as well as rounded pebbles and cobbles from the underlying river deposits.

The depth and amount of underground water are major factors controlling the locations of explosive basalt volcanoes on the eastern Snake River Plain. Large explosive complexes such as the Massacre Volcanics and the Menan Buttes formed along the southern margin of the Snake River Plain, near the ancestral Snake River and in other places where the ground was saturated with water. By contrast, such features rarely occur and are small in volume on the northern part of the eastern plain, where the water table is deep underground, in many places greater than 500 feet.

Although basaltic volcanism has dominated the eastern plain for the past several million years, a few rhyolite volcanoes have erupted in the past half million years or so, forming conspicuous landmarks that were used by pioneers as they traveled westward on the Oregon Trail.

Big Southern Butte and East Butte are domes of viscous rhyolite lava that erupted through the basalt flows. Middle Butte is a cap of basalt lava flows, believed to have been uplifted by yet another rhyolite dome that never reached the surface.

The Menan Buttes were formed when rising basaltic magma encountered water-saturated river gravels. The resulting explosions threw out tuffs, gravels and exploded lava, creating the buttes; left, Big Southern Butte is a dome of rhyolite lava that erupted through the basalt lava 300,000 years ago.

CRATERS OF THE MOON

Along the Great Rift a sea of stone extends in all directions. Waves and whirlpools of blue-black lava stand fixed in time. The Great Rift crosses the eastern Snake River Plain as a narrow zone of open cracks, fissure-fed lava flows and pyroclastic vents whose northern end is marked by the Craters of the Moon lava field.

Over the past 15,000 years, hundreds of lava flows, cinder cones and spatter cones were formed during eight major cycles of volcanism at Craters of the Moon. Radiocarbon dates of charred vegetation under the lavas and cinders suggest that each of the eight eruptive periods probably lasted only a few decades or centuries, but were separated by long, quiet intervals of a thousand years or more. About 8 cubic miles of lava and tephra were erupted at Craters of the Moon, making it one of the largest postglacial lava fields in North America.

Early twentieth-century geologists and adventurers such as Harold Stearns and Robert Limbert were among the first to write about this raw landscape and its diverse community of plants and animals. Their superlative descriptions and recommendations helped generate interest in the Craters of the Moon National Monument, established in 1924.

The Craters of the Moon lava flows have an unusually wide-ranging chemical makeup; they are much more diverse than is typical of basalts elsewhere on the plain. Craters of the Moon lavas range from highly fluid, gas-poor basalt, which created small shield volcanoes, small spatter cones and tube-fed lava flows, to pasty, gas-rich lavas that erupted violently to form large cinder cones, or poured sluggishly from open cracks to form rubbly, steep-sided lava flows. These landforms mimic those of better-known hot-spot volcanoes such as the Hawaiian Islands, where lava flows are quickly covered by tropical vegetation and are soon weathered by drenching rains. But in the arid climate of southern Idaho, soils form slowly and plants find it hard to gain a foothold. Thus, the several-thousand-year-old cones and flows seem as if they formed only yesterday.

The geologically young, voluminous lava outpourings at Craters of the Moon seem to betray the hot-spot theory for the origin of the plain. If the hot spot is now reckoned to underlie Yellowstone, then why have the most recent eruptions occurred at Craters of the Moon, 150 miles from the hot spot? Although the Craters site is the largest postglacial lava field on the Snake River Plain, some eight more lava fields have also formed in the region within the past 15,000 years. These young lava fields erupted from several volcanic rift zones, and together they cover more than 10 percent of the eastern Snake River Plain. Geologists view this as "residual volcanism," volumetrically unimportant in the grand scheme of things, but hinting at the huge amount of heat and magma that still remains under the Snake River Plain more than 6 million years after the hot spot passed beneath the area.

Are future eruptions likely here? Volcanism at Craters of the Moon spanned 15,000 years of time, and its eight eruptive periods were separated by quiet intervals averaging about 2,000 years. Craters of the Moon last erupted a little over 2,000 years ago. Could it happen again? Geologists predict it will.

Borah Peak, the tallest mountain in Idaho, is part of the Basin and Range Province — fault-block mountains that rise abruptly from the surrounding lowlands.

MOUNTAIN NEIGHBORS

The mountain ranges of southern Idaho end abruptly against the eastern Snake River Plain. The highest point in the state — 12,662-foot Borah Peak — is only 50 miles from the flat, mile-high lava plains.

The surrounding peaks are part of the Basin and Range Province, a distinctive region of mountains and valleys that are found throughout much of the western United States and northern Mexico. The region includes most of the terrain between the Rocky Mountains and the Sierra Nevada of California, where geologically young, precipitous mountains rise from encircling aprons of sediment that have been shed into the intervening valleys. There are more than 150 mountain ranges in the Basin and Range Province, each trending north-south and rising 3,000 to 5,000 feet above the surrounding lowlands.

Most of the ranges in southeastern Idaho are composed of uplifted sedimentary rocks deposited in ancient seas that intermittently covered parts of western North America between about 200 and 700 million years ago. Many ranges are capped by lava or volcanic tuff that was erupted much later, as the mountains began to uplift. The mountain ranges are typically asymmetrical with a steep slope on one side and a gentler one opposite due to curved faults that have uplifted one side of the mountain more than the other. The mountains "grow" by the uplifting and tilting of ancient strata, much like rows of closely spaced dominoes that have been pushed over. Weathering and erosion of the uplifted rocks then produces large amounts of sediment.

Mountain uplift and the shedding of sediments into the adjacent valleys have gone on for about the past 17 million years in the Basin and Range Province. Geological reconstructions suggest that the Earth's crust has been stretched in an east-west direction

by at least 50 percent during that time. The cause of the stretching can be attributed to upwelling convection currents in the Earth's mantle beneath a huge area of western North America. In the stretching process, cool, brittle rocks within a few miles of the Earth's surface fracture and fault into rugged mountains, but hotter rocks of the deep crust stretch like taffy.

Mountain building is an active process in Idaho today, as evidenced by numerous historical earthquakes to the north and south of the eastern Snake River Plain. California's frequent tremors notwithstanding, two of the largest earthquakes recorded within the United States occurred at Hebgen Lake, Montana, in 1959 and near Borah Peak, Idaho, in 1983. Both tremors were greater than magnitude 7 on the Richter scale. Both were located within a zone of frequent earthquakes that extends from eastern-central Utah into eastern Idaho and Montana, marking the active eastern edge of the Basin and Range Province. During both earthquakes, the ground was broken along mountain fronts, indicating uplift of a mountain range and downthrow of its adjacent valley. Thus, the forces that have been pulling apart the crust of the western United States during the past 17 million years continue today, causing renewed uplift of jagged, youthful mountains that would otherwise be quickly worn away by erosion.

The Borah Peak earthquake of 1983 was part of the fault-block mountain-building process that has lifted the Lost River Range. While the focus — the origin of the earthquake — was located beneath the valley, a fault scarp showing the vertical displacement of the quake cut across the lower flank of the mountain range.

SOILS OF THE PLAIN

During the lengthy intervals between basaltic lava outpourings on the eastern plain, large tracts of barren lava were blanketed by thick accumulations of wind-blown silt known as loess. Deposits continue to accumulate today during wind storms. Microscopic examination of loess shows that it contains little volcanic material and is mostly made of components derived from older lake beds to the west. The well-drained, fertile soils of southern and eastern Idaho thus are not formed directly by the weathering of the underlying volcanic rocks, but have been blown onto the Snake River Plain by winds during the past several million years.

In places, the wind velocities and the supply of sandy material are great enough to produce large sand dunes. The most impressive sand dunes on the plain occur at the Bruneau Dunes State Park in southwestern Idaho and in eastern Idaho near the town of St. Anthony, where active and vegetated dunes partially cover rhyolite ash-flow tuffs and basalt lava flows of the Juniper Buttes.

Wind has been the single-most important element in soil formation on the Snake River Plain. The Bruneau Sand Dunes, above, were created by wind-borne sand accumulated and trapped in an abandoned meander of the Snake River. Right, wind-blown silt known as loess blankets much of the Snake River Plain.

THE YELLOWSTONE COUNTRY

Although the early period of explosive rhyolitic volcanism has ended on the plain and those volcanic centers are partly or entirely covered by younger basaltic lava flows, relatively recent rhyolitic volcanism is still quite evident in the broad region of thermal features on the Yellowstone Plateau. Island Park in eastern Idaho is geologically a transition between Yellowstone and the low-lying terrain of the eastern plain. The Island Park topo-graphic basin was formed by several cycles of rhyolitic and basaltic volcanism during the last 2.1 million years. Each cycle apparently built over a period of several hundred thousand years, as rhyolite lava flows were erupted from ringlike fractures above a large and growing magma chamber. Each cycle culminated with the explosive eruption of ash-flow tuffs and widespread airfall ash. Each catastrophic eruption probably lasted only a few days but produced 70 to 600 cubic miles of ash-flow tuff. By comparison, the total volume of material emitted from Mount St. Helens in 1980 was less than 1 cubic mile.

The largest of the ash-flow sheets — the Huckleberry Ridge Tuff — was erupted about 2.1 million years ago from a caldera that encompassed most of the Yellowstone–Island Park region. A segment of that caldera still remains as Big Bend Ridge, a curved ridge that forms the southern and western boundaries of the Island Park basin. During the second cycle of rhyolitic volcanism, the Mesa Falls Tuff (70 cubic miles in volume) was erupted from the Henry's Fork caldera, a smaller collapse depression nested within the first-cycle caldera. Within the Henry's Fork caldera, the Upper and Lower Mesa Falls of the Henry's Fork cascade over cliffs of the densely welded Mesa Falls Tuff.

The last climactic rhyolite volcanism in the Yellowstone–Island Park area produced the Lava Creek Tuff, which erupted about 600,000 years ago from the Yellowstone caldera in what is now the center of the national park. The tuff has an estimated volume of nearly 250 cubic miles, and covers most of Yellowstone National Park and the Island Park basin. After eruption of the Lava Creek Tuff, large rhyolite lava flows issued from ringlike fractures of the Yellowstone caldera, and several of them traveled westward, where they now form the eastern topographic rim of the Island Park basin. Thus, what has been called the "Island Park caldera" is not a single caldera, but a complex feature that formed during three cycles of rhyolitic volcanism spanning the past 2.1 million years.

Opposite, at Mesa Falls the Henry's Fork of the Snake River cascades over cliffs of densely welded Mesa Falls Tuff.

YELLOWSTONE PARK

"One day," said Ferdinand V. Hayden, "the intelligent American will look upon the map and recognize with conscious pride that this place hath not its equal in all the world." Hayden, writing from the Yellowstone country in 1874, belatedly confirmed the natural wonders that trappers like John Colter had known since Jeffersonian times. In an effort to preserve its exotic features, Congress set aside much of what is now northwestern Wyoming and adjacent portions of Idaho and Montana. It was America's first national park. The rugged and geologically youthful terrain of Yellowstone is mostly of volcanic and glacial origin. The park covers the central part of the world's largest thermal area — a 1,400-square-mile, mountainous plateau with an average elevation of about 7,000 feet. The park itself, about a third of the plateau, preserves outstanding hot springs, geysers and mud pots, all surface expressions of a huge 2-million-year-old magma system overlying a hot spot in the Earth's upper mantle.

The Upper Geyser Basin of Yellowstone National Park has ideal characteristics that have created the largest concentration of active geysers in the world: permeable sand and gravel deposits overlying fractured volcanic rocks that allow the upward passage of hot fluids.

Over the past 2 million years the plateau has been the site of huge but infrequent eruptions of rhyolite lava, ash and pumice. The last eruptions of rhyolite lava occurred about 70,000 years ago.

The Yellowstone caldera is a broad, oval-shaped depression about 25 by 45 miles in size. It formed about 600,000 years ago when the last cycle of volcanism climaxed and the Lava Creek Tuff was explosively erupted. Since that time, voluminous lava flows of rhyolite obsidian have obscured the rim of the Yellowstone caldera.

At present, the Yellowstone Plateau is one of the most seismically active regions of North America. During the 15 years between 1973 and 1988, scientists recorded about 15,000 earthquakes greater than 2 on the Richter scale. In 1959, the 7.5 Hebgen Lake earthquake occurred along a fault centered about 10 miles west of Yellowstone. The largest tremor ever recorded in the Rockies, it changed the behavior of hundreds of geysers and hot springs. Earthquakes help maintain the thermal activity of the region. If not for tremors, mineral matter would quickly seal off the fault zones that channel fluids to the surface.

A large body of partly molten rock lies beneath the northeastern part of the Yellowstone caldera. The mass of hot material starts at a depth of about 6 miles, encompasses the whole caldera, and extends downward into the Earth's mantle. Called the Yellowstone plume or hot spot, it has been the source of the 1,000 cubic miles of rhyolitic ash and lava that have erupted from the caldera during the last 2 million years.

GEOTHERMAL FEATURES

The Yellowstone hot spot is also the source of heat for nearly 10,000 hot springs and several hundred geysers. The hot springs and geysers occur when rain and snowmelt soak into the ground and percolate through fractured volcanic rocks to depths of a mile or more. There the fluids are heated. Hot water returns to the surface along permeable faults. Thus, the thermal features of Yellowstone are not randomly distributed throughout the park, but tend to occur where fluid pathways are available through fractured rocks.

Geysers — saline hot springs that erupt — are short-lived and fragile features. Many have become inactive during historic time, as a result both of natural events and of man's activities. Of the 10 major geyser areas in the world, only three — in Yellowstone, Iceland, and Kamchatka — remain essentially undisturbed.

Geysers require very specific physical and thermal conditions. There must be a potent source of heat, usually magma or hot-but-solidified rocks; abundant underground water must form a deep convection system; and there must be a focused pathway (usually a major fault or fracture in the Earth's crust) for the hot fluids to rise toward the surface at temperatures close to boiling. Near the surface, the underground water and steam need a shallow, cavernous storage system. These systems must have a constricted surface opening in order to squirt fluids into the air. Finally, the storage system must be capable of recharging and reheating between eruptions.

When these conditions are not met, features other than true geysers develop. If there is no focused pathway to the surface, fluids may remain underground. If the heat source is

Heat and pressure within the Earth cause water to rise to the surface, where it erupts in a variety of forms, such as the geysers, mud pots and hot springs found in Yellowstone Park.

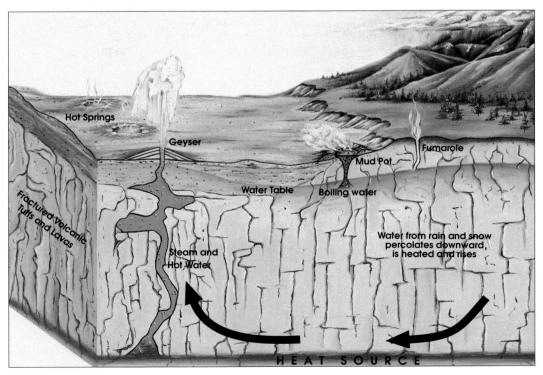

not hot enough, or if there is too much cool ground water, then warm springs may result. An example is the travertine (limestone) terraces at Mammoth Hot Springs, Wyoming. If underground temperatures are very high but there is little groundwater present, then dry steam vents (fumaroles) will form. Acid vapors from fumaroles alter the volcanic rocks into clay, forming mud pots and mud volcanoes. If the surface opening is too large or if the shallow reservoir allows free circulation, then instabilities may not develop and the hot spring may simply boil but not erupt. Such vents are called "perpetual spouters."

Although causes of geyser eruptions are complex, the phenomenon can be described generally in terms of the boiling behavior of water. Geysers are unstable saline hot springs that contain boiling or near-boiling water. Their underground plumbing systems might be compared to cavernous, vertical pipes that may extend several hundred feet underground. On the surface, the water column is boiling at atmospheric pressure and its temperature is thus about 199 degrees F. But hundreds of feet underground, the water is under considerable hydrostatic pressure. Thus, water may reach a temperature of several hundred degrees. When the water in any part of the system is heated close to its boiling point, then a very small drop in pressure will cause the water to explosively "flash" into

Above, Beehive Geyser — the timing of eruptions depends on how long it takes the "plumbing system" to be recharged by hot water and how long it takes to reheat the water to boiling temperature; left, the vivid colors of pools such as Grand Prismatic Spring result from microorganisms living in the water.

Travertine terraces at Mammoth Hot Springs.

steam. This can happen as a consequence of boiling in the upper levels of the water column, since a pressure drop is caused by rising steam bubbles. At some point in the plumbing system, flashing eventually occurs, and a mixture of steam and water is propelled out of the orifice. The eruption of fluids lowers the pressures in successively deeper levels of the system and flashing occurs as a downward-propagating chain reaction. The steam explosion causes fluids to be erupted from the reservoir.

The time between geyser eruptions depends on how long it takes the plumbing system to be recharged with hot water, the time required for this water to reheat to its boiling point, and the amount of fluid that was discharged during the previous event. Some irregularity is thus typical of most geysers, including "Old Faithful."

FERDINAND V. HAYDEN
(1829–1887)

Excitable, impulsive, a gifted fossil-hunter with a flair for self-promotion, Ferdinand Hayden was to western science what John J. Astor had been to the fur trade: its visionary thinker, its marketing genius and entrepreneur. As head of the wide-ranging United States Geological Survey of the Territories, 1867 to 1879, Hayden brought photographers, artists, scientists and surveyors to some of the West's most exotic vistas. His Snake River work was mostly confined to the headwaters in the Grand Teton region, but Hayden, a theorist, was one of the first to advance the notion that Idaho and much of the

Hayden, at left, in camp.

West were the beds of ancient freshwater lakes. His masterpiece of promotional science was an 1871 report that sold Congress on the idea of a wilderness sanctuary in the Great Geyser Basin of the Yellowstone country, now a national park. Hayden immediately understood that Yellowstone was the vast crater of a giant eruption. It was a "land of enchantment," Hayden reported. "All the brilliant feats of fairies and genies in the Arabian Nights' Entertainments are forgotten in the actual presence of such marvelous beauty."

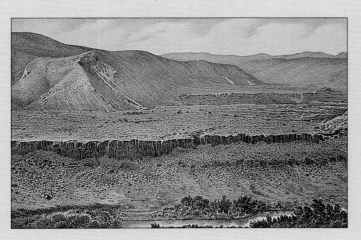

Hayden's study of the Lower Portneuf Valley, 1879.

As the sun rises over Big Southern Butte and the Great Rift, the Snake River Plain seems timeless and unchanging. But the geologic processes that created the plain continue, and the Great Rift may once again open and release rivers of molten stone.

OUR FUTURE, OUR PAST

The spectacular thermal features of Yellowstone and the Snake River country are ours to enjoy during this respite between major eruptions. What might tomorrow bring? Mark Twain spoke for geologists when he claimed that it was "extremely difficult to make predictions, especially when they concern the future." And yet we make the attempt.

The precise nature and timing of future events is surely beyond our present knowl–edge, but the violent volcanic past is a guide to some general assumptions. On the eastern plain, the Great Rift, which has been active since the Ice Age, is the most likely region of future volcanism. It is conceivable that readers of this book might one day visit the Craters of the Moon area to witness basaltic eruptions along the Great Rift. It also is quite probable that explosive eruptions will rock Yellowstone sometime during the next few hundred thousand years.

In the more distant future, the Yellowstone area may one day become a sage-covered depression much like the eastern plain. In millions or tens of millions of years, a section of southwestern Montana may have its own large caldera northeast of Yellowstone, a future volcanic area of hot springs and geysers. Although this certainly reaches beyond the span of human existence, to the geologist it seems a moment away — less than 1 percent of the Earth's history. The time span of scientific imagination knows no bounds.

In the meantime, we can refine our predictions by studying the catastrophic events that created the volcanic crescent, a harsh setting for life. Here on barren lava the plants fight for precarious footholds. Wildlife follows lost rivers, searching porous rock for marginal water sources. Humans, repelled by treacherous lava, are attracted to it as well. All inhabitants of the plain contend with the ancient forces that continue to mold our lives.

Cinquefoil

THE SNAKE RIVER PLAIN contains a wealth of natural features that attract the attention of roadside geologists. Most of the sites are easily accessible on state highways. Some of them are listed here from west to east.

Bruneau and Jarbidge River Canyons
The Bruneau and Jarbidge rivers have carved several-hundred-feet-deep canyons into the high desert plateau of Owyhee County. The canyons, carved in lava following the sudden draining of Lake Idaho, provide a cross-section of the eruptions that occurred in southwestern Idaho. *Southeast of Bruneau.*

Bruneau Dunes
At the Bruneau Dunes, wind-borne sand has been deposited in a depression created by an old meander in the Snake River. One dune is 470 feet tall. *East of Bruneau.*

Malad Gorge
The Malad River cuts through a 250-foot gorge and plunges down stairstep falls before reaching the Snake River. *West of Bliss on Interstate 84.*

Thousand Springs
At Thousand Springs, ground-water from the Snake River Aquifer bursts from the canyon walls and flows into the Snake. Much of the flow has been diverted for power production and trout farming. *East of Hagerman on Highway 30.*

Shoshone Falls
When there is enough water in the Snake River, the 212-foot-high falls are a thundering cataract. Geologists believe the falls were created during the Bonneville Flood 15,000 years ago. Today, upstream irrigation diverts water, leaving only a trickle for most of the year. *Northeast of Twin Falls.*

Shoshone Ice Caves
In these lava tubes a constant temperature of 28–33 degrees creates natural refrigeration that allows ice to remain year-round. The cave is 1,000 feet long and 40 feet high. Mammoth Cave, a one-mile-long lava tube, is nearby. *North of Shoshone on Highway 75.*

Craters of the Moon
"Craters" is the northwestern section of the Great Rift volcanic rift zone. The Craters of the Moon lava field was formed by dozens of flows and pyroclastic cones that formed during eight eruptive periods over the last 15,000 years. *West of Arco on Highway 93.*

Big Southern Butte
Rising 2,500 feet from the plain, the butte is rhyolite that erupted through the basalt flows 300,000 years ago. *Southeast of Arco.*

City of Rocks
Some of the oldest formations in the United States are found among the towering granite rocks that have been carved into a variety of interesting shapes by centuries of weathering. *South of Albion.*

Massacre Rocks
Here, the Snake River has cut through the basalt veneer, exposing older rhyolite rock beneath. Remnants of the Bonneville Flood are seen in the boulder bars within the canyon and the channels carved adjacent to the canyon on the north side. *West of American Falls on Interstate 86.*

Hells Half Acre Lava Field
This 180-square-mile lava field is radiocarbon dated to be 5,200 years old. *Between Blackfoot and Idaho Falls.*

Menan Buttes
The largest of the two buttes rises 800 feet high. Both are formed of basalt tuff and have deep craters. *Near Rexburg.*

St. Anthony Sand Dunes
Sand blown across the Snake River Plain has collected in this area of 150 square miles. *North of St. Anthony.*

Mesa Falls
At Upper and Lower Mesa Falls the Henry's Fork of the Snake River cascades over cliffs of densely welded tuff. The upper falls drops 114 feet. *Northeast of Ashton.*

Big Springs
Issuing from rhyolite lava flows, Big Springs is a major source of the Henry's Fork of the Snake River. *Eastern margin of the Island Park basin.*

Bruneau Sand Dunes

The desert is a vast world, an oceanic world,
as deep in its way and complex
and various as the sea.

— *Edward Abbey*

Desert Solitaire, 1968

A CLIMATE OF CHANGE

THE SNAKE RIVER PLAIN is one of the last discovered and least understood regions in the American West. It has been — and continues to be — a landscape of change. Over the past 50 million years, the region has been humid, frozen, marshy and arid. Its forests, lakes and grass-lands have supported a rich diversity of life, much of it extinct. Even

BY E. B. BENTLEY
AND GLENN OAKLEY

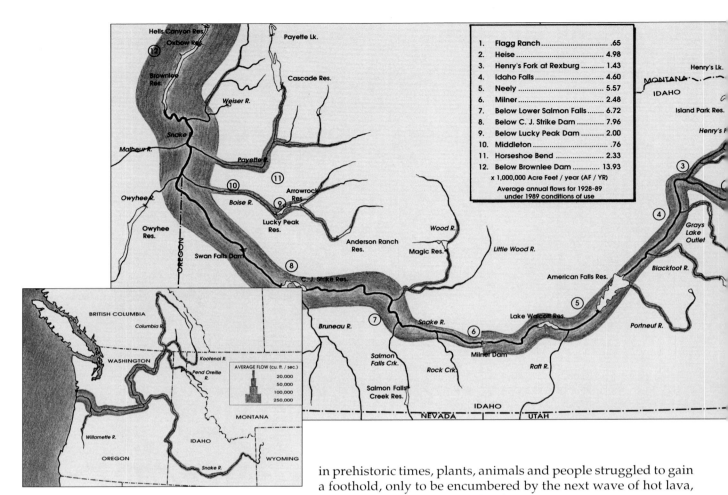

1. Flagg Ranch	.65
2. Heise	4.98
3. Henry's Fork at Rexburg	1.43
4. Idaho Falls	4.60
5. Neely	5.57
6. Milner	2.48
7. Below Lower Salmon Falls	6.72
8. Below C. J. Strike Dam	7.96
9. Below Lucky Peak Dam	2.00
10. Middleton	.76
11. Horseshoe Bend	2.33
12. Below Brownlee Dam	13.93

x 1,000,000 Acre Feet / year (AF / YR)

Average annual flows for 1928-89 under 1989 conditions of use

AVERAGE FLOW (cu. ft. / sec.)
20,000
50,000
100,000
250,000

Columbia basin; previous page, mural of the humid Carboniferous period, 300 million years ago; detail, green heron.

in prehistoric times, plants, animals and people struggled to gain a foothold, only to be encumbered by the next wave of hot lava, catastrophic flood or dust cloud.

In historic times, the ecology of the plain has changed as it has passed through the eras of Native American inhabitance, exploration and mountain men, the Oregon Trail and then the railroad. With increased settlement, water projects and agriculture expanded rapidly, leading to the growth of towns and cities.

The Snake River Plain transcends mere description; rather, it is a place to experience. To really understand the plain, one must choke on the dust, feel the biting wind and driving sand, smell the sage after a summer cloudburst, see the spray of wildflowers in the spring, measure the profound effect of a rattlesnake inches away or look in awe at the chasm of the Snake or Bruneau rivers and hear the water rushing far below. Imagine the wagon master when he stood at the rim of a canyon that slices through the lava. He could throw a rock across the chasm, but it would require three days of hot and weary travel to circumnavigate the cleft.

Today, many of the geographic details of the plain are hidden from the monotonous highway that traverses southern Idaho. In places, the plain is lonely and seems empty. But it is not. It is home to more than 35 species of mammals, 136 species of birds and 25 species of reptiles and amphibians. Carpeted with sage and grasses, the desertlike environment also supports many species of herbs and wildflowers and, along the rivers, riparian areas of willows, cottonwood and birch. The plain, moreover, is a place where people have gained a foothold and now thrive, a place where farms, ranches, small towns and urban areas all exert their own changes on the landscape.

Cities on the plain will continue to grow. Likely, some small towns will perish, just as in the past. But intense loyalty for one's sense of place will continue, manifesting itself through regional competition among schools, towns and legislative districts. Kimama

Snake River, average annual flow. Draining 109,000 square miles, the Snake carries 37 million acre-feet in an average year, more than twice the volume of the Colorado. In the diagram, left, the dams and canals divert water for irrigaton. At Milner, for example, the Snake is cut to a trickle. The Boise, Payette, and other tributaries, as well as Thousand Springs, replenish the Snake as it curves into the Columbia en route to the sea.

PACIFIC COAST RIVERS

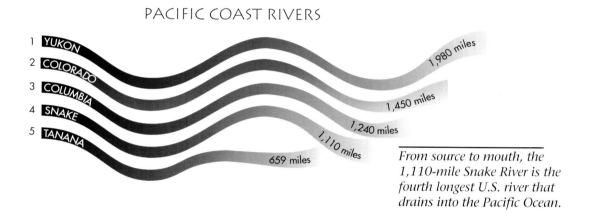

1 YUKON — 1,980 miles
2 COLORADO — 1,450 miles
3 COLUMBIA — 1,240 miles
4 SNAKE — 1,110 miles
5 TANANA — 659 miles

From source to mouth, the 1,110-mile Snake River is the fourth longest U.S. river that drains into the Pacific Ocean.

will remain Kimama and Boise will remain as that great alien city from which authority emanates. People in Shoshone will defend vociferously their place on the map along with Melba, Eden, Tiperary and Taber. The geologic, geographic and human story is told on the face of that map. It is true that the rivers, mountains and deserts are striking features, but some of the more subtle stories emerge from the place names. Glenns Ferry, Sugar City, Atomic City, Power County, Henry's Fork and many others speak directly to chapters in the human history of the plain. New names will appear and perhaps some will be lost, all adding to the story of this richly complex, constantly evolving landscape.

CHANGING CLIMATES

To understand the present Snake River Plain, we must look to the past. The sediments of the plain are a library of information concerning geologic history, changing climate and evolving biological species, all of which beg for study and interpretation.

Throughout the geologic history of the plain, lava flows have oozed over the land–scape, searing everything in their paths. Frequently the lava spilled into a river or canyon, creating a divergence and causing the water to find a different path. The result is seen in lakebed sediments now lying exposed in places such as Glenns Ferry, Bruneau and Ontario, Oregon. Adjacent highlands provided a source for materials such as silt, sand and gravel, which were deposited on the plain.

Throughout most of the plain's existence, temperatures were warmer and there was more precipitation than today. As a consequence, vegetation and animal species were markedly different. About 50 million years ago, in the subtropical Eocene Epoch, the plain supported tree palms and swampy-marshy landscapes. As time progressed, volcanic processes increased in intensity. Lava flows resulted in the relocation of streams, and ash

EVOLUTION OF THE SNAKE RIVER PLAIN

EPOCH	YEARS AGO number of years before present
Holocene	**present**
cool and dry, bison, wolf, bear, mastodon, sage, grass	
Pleistocene	**10,000**
cold and moist, camel, horse, beaver, mastodon, musk-ox, bear, bison, aboriginal humans, pines, grasses, sage	
Pliocene	**1.6 million**
formation of Snake River graben, climate becoming arid, horse, rhino, camel, mastodon, pines, grasses	
Miocene	**5.3 million**
mild and humid, active volcanoes, primitive horse, rhino, camel, redwoods, oaks	
Oligocene	**23.7 million**
warm and humid, active volcanoes, floods, mudflows	
Eocene	**36.6 million**
probably subtropical, four-toed horse, rhino, tree palms, swamp, marshland	
Paleocene	**57.8 million**
warm and humid, subtropical	
Cretaceous	**66.4 million**
formation of Idaho Batholith	

Source: Geological Society of America Standard Correlation Chart

Bruneau Canyon, about 1922.

deposits accumulated in shallow lakes and marshlands. Subtropical conditions waned and the climate became less humid.

Climate conditions through the Miocene (about 23 to 5 million years ago) resulted in forest cover that might be similar to Appalachian forests of today — mainly oak, maple, sassafras and willow. The adjacent hill slopes were forested with redwoods, and the animal population included three-toed horses, rhino, camel and saber-toothed cat. The geologic events that led to the contemporary form of the plain began during the Miocene Epoch about 15 million years ago. Increased volcanic activities to the west led to frequent ash deposits that covered and preserved the fossil remains from that epoch.

At the onset of the Pliocene, about 5.3 million years ago, the climate of the area lying between the Sierra Nevada–Cascades on the west and the Rocky Mountains on the east became increasingly more arid as more rain was blocked by the uplifted Cascade Mountains. Pine became the dominant tree species and grasses pioneered much of the plain. Consequently, the horse, mastodon, giant sloth and camel roamed the rangelands along with the carnivorous bear.

Excavations from the Hagerman Fossil Beds indicate that the Early Pliocene was more humid than today and that water was abundant. Deposits indicate vigorous plant growth in backwater swamps. Thin layers of gypsum, however, are evidence that intervening drying periods resulted in evaporative deposits. During these drier periods there was a proliferation of deciduous forest trees and attendant animal populations. There also is evidence of early camel, bear, antelope, wolf and an early horse. During this time beaver had gained a foothold and were abundant on the plain.

Near the close of the Pliocene Epoch, about 1.6 million years ago, the area experienced a general cooling trend. Conditions and events that prevailed during this period are commonly referred to as the Ice Age, or more precisely, the Pleistocene Epoch. For the following 1.6 million years, temperature and moisture conditions varied from cooler than at present to warmer than at present.

Such changes in climate also altered plant and animal populations. Scientists have determined through the study of freshwater lakes in the Great Basin that during the cool episodes, temperatures on the plain was about 8–9 degrees cooler and that precipitation was about 10 inches more than today. The mean annual temperature on the western portion of the plain would have been about 45 degrees F, similar to the conditions that presently exist in north-central British Columbia.

In the Idaho Falls area on the eastern plain, the mean annual temperature during the cold phases was about 34 degrees F, or similar to the conditions that prevail at Dawson Creek in northeastern British Columbia. The upper reaches of the plain near the Yellowstone Plateau would have been especially harsh, with temperature conditions comparable to those experienced above the Arctic Circle in Yukon Territory.

Hagerman horse

HAGERMAN FOSSIL BEDS NATIONAL MONUMENT

The Hagerman Fossil Beds National Monument is located on the west bank of the Snake River near Hagerman. The fossil beds are preserved in flood plain sediments associated with the enormous Late Pliocene "Lake Idaho" that covered much of southwestern Idaho around 3.5 million years ago. Within the monument more than 300 fossil sites are exposed in 500 feet of the sedimentary cliffs, which represent around 180,000 years of flood plain deposits. There are 105 vertebrates and 39 invertebrates known from the Hagerman site, and 44 species were first described from fossils found at this site, which is internationally famous as one of the richest known Late Pliocene fossil deposits. There are more than 160 published paleontologic references that cite specimens from the Hagerman Fossil Beds.

Two distinct habitats are represented by the upper and lower areas of the fossil-rich deposits in the monument. Fossils in the lower part of the cliffs indicate a lake or marshland habitat and include muskrat, voles, fish, frogs, water snakes, turtles and birds such as pelicans, ducks, cormorants and swans. The most abundant mammals are beaver, muskrat, otterlike animals and others associated with an aquatic environment. Pollen from hackberry, poplars and pines indicate a warmer, more humid climate than today's.

The upper or most recent habitat is a riverbank flood plain, home to ground sloths, herbivores such as the "Hagerman horse," peccaries, camels, deer, antelope and mastodon. Carnivores include bear, hyenalike dogs, saber-toothed cats, dogs and weasellike predators. The presence of burrowing rodents such as ground squirrels, pocket gophers and kangaroo rats, as well as toads, hares, shrews and lemmings are evidence of a drier, subtropical savannah condition. The animals and plants present suggest that the wetland or marshy habitat in the lower areas of the cliffs was also gradually replaced by a drier subtropical savannah at this site.

The zebralike Hagerman horse (*Equus simplicidens*), for which the monument is best known, is the Idaho State Fossil. The U.S. National Museum of Natural History (Smithsonian Institution) excavated 160 horse skulls and 20 complete skeletons from the "Horse Quarry" in a series of expeditions in the 1930s.

Smithsonian paleontologist Norman H. Boss (right) and an unidentified colleague examine fossils at the Hagerman site in 1931.

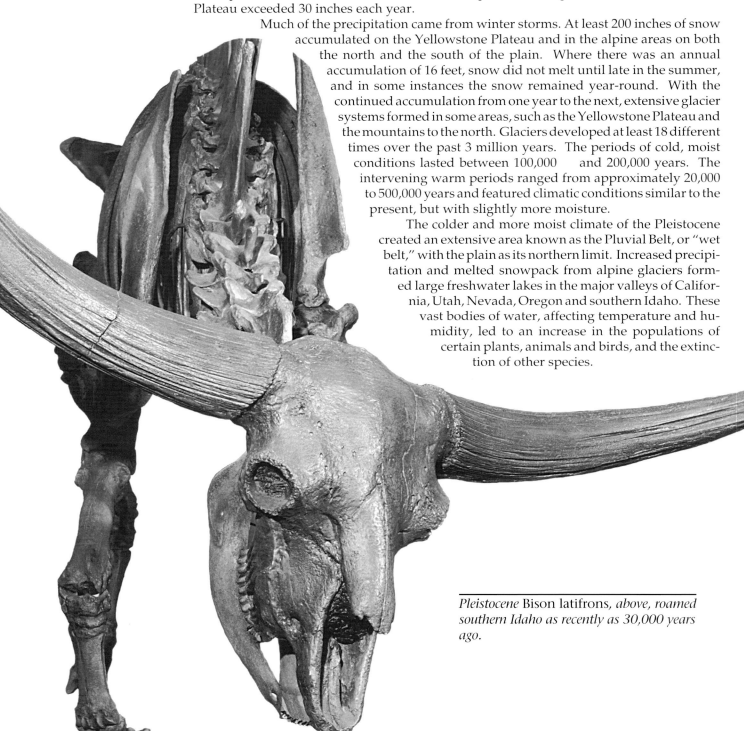

As the environment grew colder, precipitation almost doubled, much of it in the form of snow. The annual precipitation increased to about 20 inches on the western portion of the plain and to about 23 inches on the eastern portion. Precipitation on the Yellowstone Plateau exceeded 30 inches each year.

Much of the precipitation came from winter storms. At least 200 inches of snow accumulated on the Yellowstone Plateau and in the alpine areas on both the north and the south of the plain. Where there was an annual accumulation of 16 feet, snow did not melt until late in the summer, and in some instances the snow remained year-round. With the continued accumulation from one year to the next, extensive glacier systems formed in some areas, such as the Yellowstone Plateau and the mountains to the north. Glaciers developed at least 18 different times over the past 3 million years. The periods of cold, moist conditions lasted between 100,000 and 200,000 years. The intervening warm periods ranged from approximately 20,000 to 500,000 years and featured climatic conditions similar to the present, but with slightly more moisture.

The colder and more moist climate of the Pleistocene created an extensive area known as the Pluvial Belt, or "wet belt," with the plain as its northern limit. Increased precipitation and melted snowpack from alpine glaciers formed large freshwater lakes in the major valleys of California, Utah, Nevada, Oregon and southern Idaho. These vast bodies of water, affecting temperature and humidity, led to an increase in the populations of certain plants, animals and birds, and the extinction of other species.

Pleistocene Bison latifrons, *above, roamed southern Idaho as recently as 30,000 years ago.*

Rhabdofario lacustris, *an ancient relative of the rainbow-redband trout, inhabited enormous Lake Idaho during the Pliocene Epoch.*

There were a great many more small streams, lakes and marshlands on the plain. This riparian vegetation provided a habitat for increased populations of fur-bearing animals, waterfowl and even the larger grazing animals. The plain was a significant habitat for nesting birds, particularly waterfowl. While there are now about 6 million migrating waterfowl passing through the plain each year, the wetter pluvial environment may have supported twice that population. Many nested on the plain and remained throughout the year. Subarctic nesting areas froze. Some mammals were forced to adapt or perish. Some Pliocene species — such as the mastodon, beaver and musk-ox, along with predators like the dire wolf, short-faced bear, saber-toothed cat and giant lion — adjusted to the cooler, wetter clime of the Pleistocene. The giant ground sloth, mammoth and camel also adapted to cooler conditions. Paleontologists have discovered fossil evidence of these species in excavations from Weiser to American Falls. Several of these animals, such as the sloth, fed upon the leaves of deciduous trees and shrubs.

As the icy grip of the Pleistocene Epoch started to relax, large herds of bison, ante-lope, bighorn sheep, elk and deer grazed or browsed on the lush vegetation. Solitary mastodons shredded and ate the bark from aspen and cottonwood, and the plain was a cacophony of ducks, geese and gulls. Snow and ice melted at an ever increasing rate, with freshwater lakes reaching higher and higher levels. About 15,000 years ago Lake Bonneville (pre–Salt Lake) spilled over Red Rock Pass near Pocatello and quickly cut downward approximately 325 feet. The resultant deluge of water swept down the Marsh Creek Valley into the Portneuf and then into the Snake. The flood that swept westward along the course of the Snake resulted in phenomenal changes to the landscape, many of which are visible today.

The Bonneville Flood occurred near the end of the Pleistocene ice age. The climate grew milder and by about 11,000 years ago significant changes took place in plant and animal populations on the plain. Increased evaporation, decreased snowpack and less precipitation resulted in a drier climate. Soil profiles indicate that blowing dust was a normal condition almost every day. Following the retreat of glacial conditions, many lakes, swamps, small streams and gullies dried up. All of those served as sources for silt that could be picked up by prevailing winds. During this phase, winds shifted from a southwesterly direction to a prevailing northwesterly source, removing soil from the Columbia Basin, the Umatilla

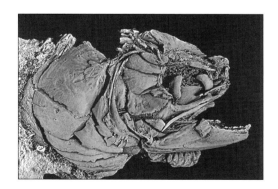

Fossils of the ancient sockeye salmon, Oncorhynchus salax, *have been found in the Glenns Ferry Formation west of Hagerman.*

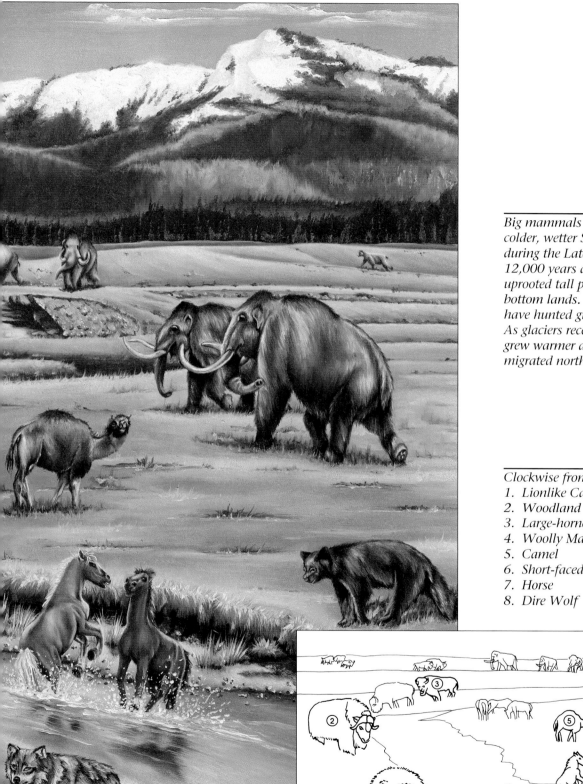

Big mammals hunted and grazed a colder, wetter Snake River Plain during the Late Pliestocene, about 12,000 years ago. Mammoths uprooted tall plants in the marshy bottom lands. Saber-toothed cats may have hunted ground sloths and bison. As glaciers receded and the climate grew warmer and drier, some species migrated north.

Clockwise from lower left:
1. Lionlike Cat
2. Woodland Musk-ox
3. Large-horned Bison
4. Woolly Mammoth
5. Camel
6. Short-faced Bear
7. Horse
8. Dire Wolf

Plateau of north-central Oregon and the western Snake River Plain near Ontario and Vale, Oregon.

Sediment removed from these areas fell like a blanket over almost every crack and crevass on the plain, accumulating to great depths in some sites and even filling subsurface lava tubes that had no apparent opening to the surface. For at least 4,000–5,000 years, great choking clouds of dust swept eastward to pile up in such areas as the Raft River Valley, Bruneau Valley, Big Lost River Sinks and St. Anthony Sand Dunes.

THE CATASTROPHIC BONNEVILLE FLOOD
AN ECOLOGICAL SCOURING OF THE SNAKE RIVER PLAIN

The catastrophic Bonneville Flood swept across the Snake River Plain about 15,000 years ago when Lake Bonneville, of which the Great Salt Lake is a relic, overtopped and eroded a natural dam at Red Rock Pass near Pocatello. The Bonneville Flood was the second largest known flood in the world. Maximum discharge occurred for eight weeks, with lower flow levels continuing much longer. Shoshone Falls, the dry cataracts below it, numerous plunge-pool alcoves and fields of rounded basalt boulders, "melon gravel," are all remnants of the flood.

A scouring flow of this magnitude had dramatic ecologic consequences. From Shoshone Falls to Bliss, the flood (around 3,030 feet) inundated the present-day two-story Snake River canyon rim to rim. In addition to removing large deposits of clay, it stripped vegetation and flooded salmon spawning habitat. Like a gigantic flushing flow, it left a morainelike alluvium over most of the canyon's first-story plateau. The canyon bench was subsequently invaded and colonized by hackberry, birch, willows and the other riparian vegetation that today dominates the river corridor. Undoubtedly, instream ecology was chaotic for years as silts and gravels were sorted as the flows decreased.

Though it was a catastrophic event from a hydrologic and geologic perspective, the rare deciduous riparian forest formations that exist along the middle Snake River riparian corridor survive in the alluvial soils left by the flood. Thus, perhaps in the long-term view, the flood produced ecological pluses as well as the devastating impact it wrought as it descended the Snake River thousands of years ago.

Lake Bonneville and present-day Great Salt Lake; below, melon gravel flood debris.

At the close of the Pleistocene, some plants became sparse or disappeared completely while others survived only in protected areas. A vigorous sea of grasses blanketed the plain. Characteristic riparian vegetation like cottonwood, birch and willows grew on the stream banks and in river canyons. Even the intermittent stream channels supported a significant growth of trees and brush because groundwater was close to the surface. Riparian vegetation provided food and refuge for browsers and carnivores alike.

About 6,250 to 3,000 years ago, subsequent to the warming trend, the climate once again became cooler and moist. During this time there was a general expansion of pine forests on the upland areas along the margins of the plain. About 3,000 years ago, the climate warmed somewhat and shadscale and sage steppe predominated. Scholars speculate about how today's processes and conditions compare to those of the past. Most agree that with the exception of a brief period from 250 to 300 years ago, it is cooler and more moist now than at any time since 9,000 years ago.

Sand dunes near St. Anthony partially cover the lava flows of the Juniper Buttes.

THE HUMAN IMPACT

Even a cursory look at the plain indicates that plant and animal populations are not the same as they were just 150 years ago. It is hard to know how much of an impact the Indians had on the native flora and fauna. Early explorers observed a landscape quite different than what is evident today. When the first Euro-Americans arrived in 1810, the beginning of the mountain man era on the upper Snake, observers described some areas of the plain as a landscape of grasses shoulder high to a horse, with brush and trees along even the intermittent streams and a riot of bird life, grazing animals and attendant carnivores. Trappers reported bison, antelope, sheep, elk, deer and beaver. Carnivores included grizzly bear, wolves, coyote, lynx and cougar.

In 1834 trapper Osborne Russell wrote that he encountered dense trees and brush in the vicinity of present-day American Falls and Pocatello. Likewise, Captain Benjamin L.E. Bonneville, crossing the plain in the 1830s, described an ocean of grass with great herds of bison. Artist Ralph Blakelock gives us a rare view in an 1871 painting titled *Indian Encampment Along the Snake* that depicts the upper Snake as a heavily forested landscape.

Russell complained that by 1843 life was not worth living on the plain because the beaver were gone, along with the bison, grizzly, bighorn sheep, elk and deer. Clearly, massive changes were taking place, and the time-frame for change coin-

Ralph Blakelock's Indian Encampment Along the Snake *(1871); inset; a contemporary view of the upper Snake.*

Oregon Trail wagon ruts west of Boise; below, federal trappers earned bounties for coyotes, badgers and other Snake River predators.

cided with the arrival of Euro-Americans. The lifestyle of the mountain man, not changes in the physical environment, led to major depletion of animal populations.

Armed only with traps and muzzle-loading rifles, a small number of people hunted out an ecosystem in a few brief years. Even as the animals disappeared, there were few roars of outrage. Quite the contrary. Many of the trappers of that era took great pride in taming the land and wresting the country away from the animals and the Indians.

But the end of the trapping era did not mean an end to ecological change on the plain. Another event, one of the most massive spontaneous migrations in history, was beginning. In the summer of 1836, a wagon train from St. Louis consisting of about 40 people and 20 wagons was the first to travel west of the Green River in Wyoming, thus opening the Oregon Trail to a steady roll of wagons across the plain. Diarists wrote that the dust was axle deep and choked animals and people alike. To make matters worse, there was little game, little water, and most importantly, almost no grass or feed for the animals. The grasses that had proliferated were increasingly scarce. Eventually, mules and oxen, grazing a wide strip across southern Idaho, ate most of the meager grasses and destroyed much of the vegetative growth on the plain.

Over a brief period of time people and the new technologies they brought with them had changed the ecosystem. The migrants moved on, not even looking back, and by the time the first farmers and ranchers settled, the plain had been altered, never to return to its pre-Euro-American form.

The eyes of most migrants were riveted westward — very few became permanent residents. Mormon farmers had established permanent town sites in the valleys to the south and east, and river crossings had become focal points. Boise had grown as a supply center for nearby mines. But the real surge of activity on most of the plain came with the arrival of the railroad. Soon after, settlements bloomed as people from all over the world came to start a new life. Like the mountain men and Oregon Trail migrants before them, their new technologies would change the ecology of the plain. Visionaries like surveyor

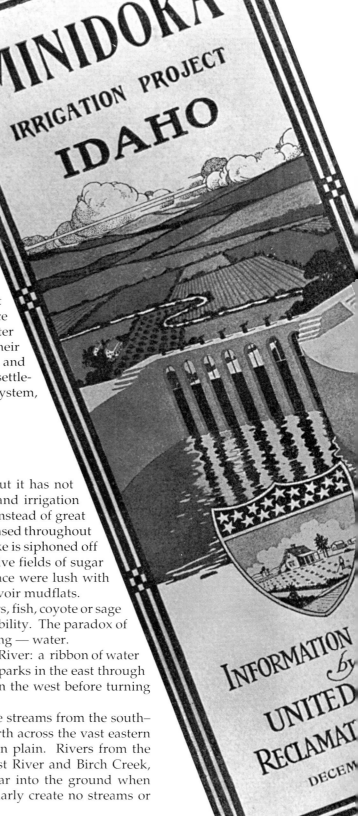

MINIDOKA
IRRIGATION PROJECT
IDAHO

INFORMATION *by*
UNITED
RECLAMAT
DECEM

Land brochure,
1909

Israel Russell and the Mormon settlers realized that water had a greater magnetism than even gold. The government was interested in it, investors were eager to embrace it, and farmers would suffer great personal hardship for the opportunity to control its flow.

In a span of approximately 40 years, from 1880 to 1920, the force of the Snake River was harnessed and agriculture came of age. With irrigation, the plain grew crops that the railroad carried to distant markets. Ditches, roads and power lines crisscrossed the desert. Roads grew so fast there was not even enough written history to provide names for them. The "old families" were last year's homesteaders, and roads were numbered according to the irrigation laterals or the township and range surveys.

By 1920, the plain supported 13 irrigation districts and at least 40 water companies. Thus, where Oregon Trail migrants once dismissed the Snake River Plain as endless heat and dust, later generations, fueled by water, began a new way of life. Their encounter with the arid landscape was paradoxical — utilitarian and exploitative. Irrigation was a double-edged sword. It brought settlement and prosperity to the region, but it also disrupted the ecosystem, forever changing life on the plain.

CORRIDORS OF LIFE

Irrigation has made parts of the Snake River Plain bloom, but it has not increased the *quantity* of water in this semidesert. Reservoirs and irrigation pipes have simply shifted the water around in time and place. Instead of great deluges in spring, the river is trapped by a series of dams and released throughout the summer. Instead of thundering over Shoshone Falls, the Snake is siphoned off in canals. Plateaus once carpeted by sagebrush are now expansive fields of sugar beets, russet potatoes and mint. Conversely, riverbanks that once were lush with wild rye, cattails, reeds and cottonwood forests are barren reservoir mudflats.

All living things on the Snake River Plain — whether farmers, fish, coyote or sage grouse — depend on water and prosper or perish with its availability. The paradox of the Snake River Plain is defined by that in which it is most lacking — water.

The single-most significant feature of the plain is the Snake River: a ribbon of water flowing from the mountains of Yellowstone and Teton national parks in the east through farmlands and the canyons of the Birds of Prey Natural Area in the west before turning north toward its confluence with the Columbia.

The Snake hugs the southern boundary of the plain. While streams from the southern mountains flow into the Snake, no rivers enter from the north across the vast eastern plain until the Little Wood flows in on the edge of the western plain. Rivers from the northern mountain ranges — the Little Lost River, the Big Lost River and Birch Creek, draining the highest mountain ranges in the state — disappear into the ground when they enter the plain. Rainfall and snowmelt on the plain similarly create no streams or

The riparian corridor of the upper Snake supports one of the West's largest cottonwood forests; below, Snake River otter; right, moose in cottonwoods along the South Fork of the Snake.

ponds. The rivers and water are absorbed into the porous lavas that cover the region to depths of several hundred feet.

The lost rivers contribute to the vast groundwater basin known as the Snake River Aquifer. It is one of the largest and most heavily used aquifers in America. But until wells were drilled into the aquifer in the 1940s, the Snake River itself was virtually the sole source of abundant water on much of the plain. Today, animals and people alike cluster along this ribbon of water and its tributaries located at the eastern and western margins of the plain: the Henry's Fork, Portneuf, Teton and Raft rivers in the east and the Owyhee, Bruneau, Jarbidge, Wood, Malad, Boise, Weiser and Payette in the west.

In the headwaters region of eastern Idaho, streams, lakes and rivers are relatively abundant. The U.S. Fish and Wildlife Service has identified the cottonwood riparian environment along the South Fork of the Snake River through Swan Valley as the most important ecological zone in the state. It is one of the largest remaining riparian cottonwood forests in the West, and cottonwood forests, it turns out, are the richest biological zones in the West.

The cottonwood forests receive year-round use by wildlife. In spring the cotton–woods are nesting areas for a myriad of neotropical birds — vireos, tanagers, warblers, northern orioles and finches — that winter in Central and South America but spend the spring and summer in the Northern Hemisphere. Bald eagles also nest in the cottonwoods, building large nests in the top half of the trees. Their smaller counterparts, osprey, build their nests at the very tops of trees. Both raptors prey on the trout and whitefish in the South Fork.

The eagles born on the South Fork tend to disperse to other regions after fledging. However, in winter the cottonwood forests see an influx of more bald eagles from Montana and western Canada. Their wintering population varies from a few dozen to more than 100, depending on weather and prey base conditions. Frequently the South Fork freezes over in the winter, forcing the eagles to hunt elsewhere. At other times an opportunity for easy prey changes their hunting patterns. In 1992–93, for example, an abundance of blacktail jackrabbits on portions of the eastern plain drew the eagles away from the river corridor.

Moose, deer, elk and their predators — black bear and cougar — live in the cotton–wood forest year-round, but the river corridor is especially important in late winter and early spring when deep snows in the higher country force the herds into the river valleys. Some animals, like the beaver and river otter, are naturally tied to the river corridor.

Riparian areas — the land immediately adjacent to streams and rivers — are the most critical habitat in virtually every ecosystem in the West. The cottonwood forests along the South Fork of the Snake are especially critical today because of their rarity. Farms and homes along the river have altered the ecology of the cottonwood forests or removed the forest altogether.

The dams that provide irrigation, hydropower and flood protection also inundate acres of cottonwood river bottom, turning miles of wildlife-rich riparian habitat into reservoirs that become sterile mudflats when the water is removed by the end of the

irrigation season. But the damage does not stop there. By ending the natural cycle of spring flooding — through which the cottonwood forests evolved over millions of years — dams have wrought damage to even the downstream stretches of river.

Cottonwoods require flooding that creates new river bars of sand, silt and gravel, where the cottony seeds, which fall from the trees in late spring like snowstorms, settle and sprout. They need the rivers to carve new channels, creating new habitat. A report jointly prepared by the U.S. Forest Service and the Bureau of Land Management states, "The number of young trees being established on gravel bars and disturbed areas is insufficient because the Palisades Dam has reduced the amount of streambed sediments and major stream channel shifting and deposition, which are needed to form new cottonwood."

BRANCHES OF THE SNAKE

Draining 109,000 square miles, an area about the size of Nevada, the Snake River is fed by a vast network of more than 130 rivers and creeks. Many of the Snake's tributaries are renowned for whitewater sports, fishing or scenic beauty. Some of these noted rivers on the Snake River Plain include:

❑ **Boise River** — Impounded by Lucky Peak Dam 10 miles east of Boise, the river feeds irrigation canals and flows into the Snake at the Oregon border. The South Fork features blue-ribbon trout fishing and a popular whitewater raft and kayak run. Much of the North Fork is a defacto wilderness river that runs through steep terrain.

❑ **Payette River** — Draining a vast area from the Sawtooth Mountains to Payette Lake, the river enters the Snake at the Oregon border near the town of Payette. The emerald green South Fork has Class IV rapids and features numerous hot springs that flow from bedrock into the river. The North Fork is famous for its Class V rapids.

❑ **Henry's Fork** — Famed among flyfishermen for its lunker trout, the river has two of the most dramatic waterfalls in the region — Upper and Lower Mesa Falls.

❑ **South Fork of the Snake** — Another popular fishing river, the South Fork yields fine catches of cutthroat and brown trout.

❑ **Bruneau and Jarbidge** — These high desert rivers in southwest Idaho cut canyons through red rhyolite and black basalt. Kayakers and rafters float the rivers during the brief spring runoff.

Opening of power plant at Swan Falls, 1901.

Some of the cottonwood stands along the South Fork sprouted in the late 1700s and are nearing the end of their lives. And researchers have found that most cottonwoods along the South Fork were established after large floods. Overgrazing by cattle in riparian zones also poses a threat to cottonwood regeneration. Biologists are currently exploring options to maintain the cottonwoods, including simulating an occasional low-level flood and removing cattle from the river corridor.

The upper tributaries of the Snake have also become the primary wintering ground for the rare trumpeter swan — at an average weight of 24 pounds, one of the largest birds in North America. Once relatively common throughout western North America, the trumpeter swan was considered extirpated from the lower 48 until a population was discovered at Red Rocks Lakes in extreme southwestern Montana. That population was protected and expanded to Yellowstone and the Henry's Fork. In the relative isolation of the Greater Yellowstone ecosystem the swan population grew well — too well.

In recent years the burgeoning population of wintering swans has overgrazed the vegetation in the rivers. And since water flows on the Snake are tapped heavily for irrigation, agricultural uses can take precedence over the needs of swans and other wildlife. Such was the case in 1989 when low water and freezing temperatures combined to trap the swans' food beneath a layer of ice. Water purchased from the irrigators was eventually released to save the swans from starvation.

With the realization that too many swans in one place increased their vulnerability, biologists with the U. S. Fish and Wildlife Service and the Idaho Fish and Game Department began capturing and transplanting swans to Seedskadee National Wildlife Refuge in Wyoming, Summer Lake in south-central Oregon and Bruneau Dunes along the Snake River in southwestern Idaho.

Fish ladder added to the dam at Swan Falls, 1922.

While much of the Snake's uppermost stretches is a wildlife haven, the river does not flow far before it is drained, diverted, dammed and polluted. The wildlife that remains lives in an unstable and potentially deadly environment. Eleven dams impound the Snake River, creating a series of reservoirs for irrigation of the surrounding desert. The river is, in fact, literally cut in half at Milner Dam, near the town of Burley, where the flow is diverted into the Northside and Southside canals. Frequently the only water left in the river below Milner is that which leaks through cracks in the structure.

The river is resurrected, bit by bit, several miles downstream from Milner Dam as irrigation runoff and springs from the Snake River Aquifer flow into the channel. In the Hagerman Valley, where the Thousand Springs flow from cliffs into the river, the aquifer escapes its subterranean flow. Downstream from the Hagerman Valley, the first tributary from the north, the Wood River, flows into the Snake. And as the Snake flows northwest across the western plain, more rivers enter from the north and south. The rejuvenated Snake River below Milner Dam, however, is held back by a series of dams that have dramatically changed the ecology of the river. The natural cycle of spring flooding has largely been halted, and without floods the Snake has been unable to scour its banks and deposit and move sand. Thus, sandbars are now rare along the Snake, replaced by mud. Higher water temperatures are created by the large, shallow reservoirs absorbing the sun's heat. The warm reservoir water released from the dams is then infused with heavy concentrations of waste from the Hagerman Valley's trout farms, dairies, municipal and residential sewage systems, farms and ranches. The combination of warm water and waste has created algae blooms that in turn use up the limited amount of dissolved oxygen in the water, killing fish by literally suffocating them in the water.

The most obvious impact of dams on wildlife has been to block the passage of fish. Salmon once traveled up the Snake River and its tributaries as far as Shoshone Falls to spawn. As early as 1895 an irrigation dam constructed on the Bruneau River blocked salmon passage up that desert river drainage. Swan Falls Dam, built in 1901, was the first dam to halt salmon on the Snake River. Although it had a fish ladder, the ladder was poorly designed. Salmon that had migrated as juveniles to the Pacific Ocean and returned as adults piled up against the concrete walls and died by the thousands.

Sixty years later the scene was repeated when Brownlee, Oxbow and Hells Canyon dams were built without any fish passage facilities. Juveniles were also unable to successfully navigate the extensive slackwater of these reservoirs in their outward journey to the sea. Attempts to trap and transport juvenile and adult salmon around the dams failed and were abandoned. The fall Chinook run alone numbered more than 10,000 above Hells Canyon.

Trumpeter swans in the morning mist.

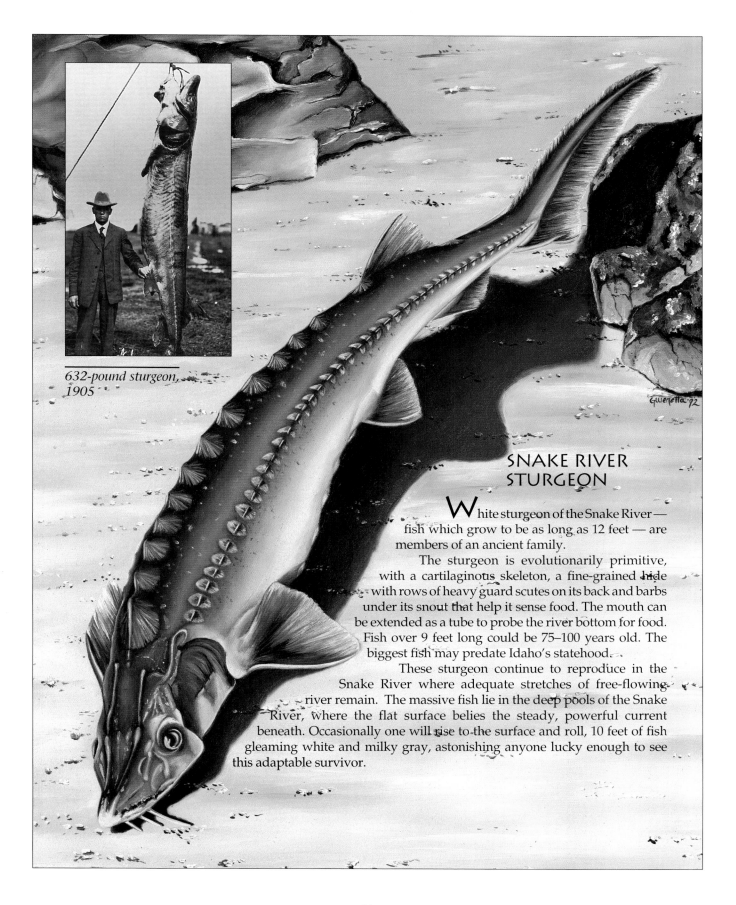

632-pound sturgeon, 1905

SNAKE RIVER STURGEON

White sturgeon of the Snake River — fish which grow to be as long as 12 feet — are members of an ancient family.

The sturgeon is evolutionarily primitive, with a cartilaginous skeleton, a fine-grained hide with rows of heavy guard scutes on its back and barbs under its snout that help it sense food. The mouth can be extended as a tube to probe the river bottom for food. Fish over 9 feet long could be 75–100 years old. The biggest fish may predate Idaho's statehood.

These sturgeon continue to reproduce in the Snake River where adequate stretches of free-flowing river remain. The massive fish lie in the deep pools of the Snake River, where the flat surface belies the steady, powerful current beneath. Occasionally one will rise to the surface and roll, 10 feet of fish gleaming white and milky gray, astonishing anyone lucky enough to see this adaptable survivor.

From 1955 to 1968, the construction of these three hydropower dams eliminated salmon runs on the rest of the Snake, Boise, Owyhee, Payette and Weiser rivers, and countless smaller tributaries. The dams similarly ended the runs of two other lesser-known species: the Pacific lamprey and the white sturgeon.

Lamprey, like salmon, swam from the ocean up into the Snake River where they would spawn and then die. Lamprey migrated up the river from May through September,

MOLLUSKS OF THE MIDDLE SNAKE RIVER

The middle Snake River is rich in aquatic mollusks, with 43 native species, including 28 snails and 15 clams. There are two introduced snails found in the river and its tributaries, and at least three more that inhabit thermal plumes at warm-water aquaculture facilities. There is only one introduced clam in the sub-basin, which at present is restricted to an area near the head of C.J. Strike Reservoir.

There is a very rich variety of mollusks in western rivers, but less than half the species that were present in the Late Pliocene Lake Idaho. Nearly all of the species in the ancient lake became extinct when it drained; however, a few cold-water species survived as relics in the middle Snake River. Of the approximately 18 species in this cold-water group half either are presently listed as endangered or threatened or are candidate species, and several others are in decline.

Poor water quality, the deposition of a blanket of organic-rich sediment in the past several decades, and the loss of habitat to impoundment have contributed to a decline in sensitive species. At present there is a transformation from a predominance of species preferring fast, cold water to those which are characteristic of warm, shallow lakes.

The ecological crash of the middle Snake River is well reflected in the mollusks. Among the local casualties are the large mussel *Margaritifera* which served as one of the food staples for Indians. This species is no longer present in the middle Snake, though old shells litter the banks. Other species, such as a number of snails, are now found only as shells. *Below, the once unique mollusk fauna of the middle Snake River, from left to right: Idaho spring snail, Snake River physa snail, Banbury Springs limpet, Bliss Rapids snail and Utah valvata snail.*

THE WHITE PELICAN

White pelicans are large birds, reaching 70 inches in length with a wingspan of up to 108 inches. They are dramatic birds, with white bodies and black feathers along their outer-wing margins. The feet and pouched beak are yellow. White pelicans nest in colonies on the ground and lay two eggs in a large nest mound. Upon hatching, the young lack feathers and must be protected from the sun. White pelicans are not solitary predators; they work together to surround fish, which they then scoop into their large beaks. Like brown pelicans, white pelicans have suffered from insecticides.

Although many early expeditions reported white pelicans in the West, the birds were not commonly found in the middle Snake River until the mid-1980s when some biologists suspect that lowered water levels in the Great Salt Lake displaced the birds. White pelicans nest as far north as central Canada, and many winter in California and Mexico, although there is a large year-round resident population in the middle Snake River. They are frequently seen in large flocks and often fly in a characteristic "V" or linear formation. They rarely make sounds, aside from a groaning sound during nesting.

although spawning did not occur until the next March or April. The females laid 30,000–100,000 eggs, which hatched within two or three weeks. After hatching, the larval lamprey, called ammocoetes, dug into the river bottom where they remained for five or more years. While in the river sediments, the ammocoetes fed on tiny desmids and diatoms. After this period the transformation from larvae to adult was complete, and they migrated to the ocean where they became parasitic. These adults attached themselves to the undersides of fish, siphoning off their blood. The ammocoetes were apparently a favorite food of the sturgeon, which would churn the river floor in search of the larvae.

Unlike the salmon and lamprey, journeys to the sea are not a life requirement for the sturgeon. They continue to live and reproduce in remnant populations in the 20 percent of the Snake that has been untouched by dams. Their population and condition is currently being evaluated.

The sturgeon is an ancient fish. Like the shark, it has cartilige, not bone. Growing to 12 feet and more, the sturgeon prowls the river bottom, feeding on mollusks, dead fish and other creatures that drift down. The Idaho record for sturgeon was a 1,500-pound fish caught in 1911 below Upper Salmon Falls. Early settlers along the Snake River caught sturgeon on set lines of heavy manila rope strung across the river and loaded them onto horse-drawn wagons, their tails hanging off the back. Commercial sturgeon fishing was common along the Columbia River and, to a lesser extent, along the Snake River in the late 19th century and early 20th century. Tom Ledbetter, who fished the Snake River near Hammett in the early 1900s, caught as many as 115 sturgeon in one year. The state rod and reel record is 364 pounds caught by Glen Howard in 1956.

The large sturgeon that still swim the Snake River probably have made several trips to the ocean and began their lives feeding on the tons of salmon carcasses that followed the spawning. Sturgeon over 9 feet may be 75–100 years old, preceding the dams that now confine them. Biologists suspect sturgeon born in recent years may never attain the size of their ancestors, since dams now block the nutrient-rich ocean migrations and the yearly influx of salmon.

Cutthroat trout and red band trout have largely been eliminated from the middle Snake by dams and pollution. Rainbow trout continue to reproduce in the middle Snake.

WHEN THE DESERT BLOOMS . . .

Blazing star

Prickly pear cactus

Desert evening primrose

Syringa

Dwarf monkeyflower

Lichen

Early summer sees wildflowers growing in profusion on the Snake River Plain. Scabland penstemon, desert parsley and cinquefoil blossom wherever a small amount of soil builds up in cracks on the lava's surface. In the deep cracks, Idaho's state flower — the syringa — grows. Buckwheat, blazing star and purple carpets of dwarf monkeyflowers cover otherwise barren cinder cones. Isolated pockets of more mature soil support rubber rabbitbrush, junipers and the gnarled limber pine. As those plants die, each one contributes its decaying roots, leaves and branches to the formation of a deep humus — the preferred habitat of sagebrush.

Other native fish species include mountain whitefish, peamouth, northern squawfish, redside shiner, speckled dace, mottled sculpin, the endemic Shoshone sculpin and largescale sucker.

Many of the other fish species now swimming the Snake were introduced by individuals or wildlife agencies. Brown and brook trout are found in the upper Snake River as well as western tributaries, the Little Wood, Boise and Payette rivers. In the late 1800s William Ridenbaugh introduced largemouth bass and perch into the Snake. Smallmouth bass apparently worked their way to the Snake from the Yakima River in Washington, where they were introduced in 1925. Since then wildlife agencies have released small–mouth directly into the river. And the much-maligned carp was introduced in 1880–1890 by the U.S. Fish Commission. Other species in the Snake include largemouth bass, channel catfish, black bullhead, black crappie, bluegill and pumpkinseed sunfish. Crappie and channel catfish now provide significant fisheries in free-flowing and impounded reaches of the Snake.

The middle Snake is also home to five mollusk species classified as threatened or endangered. The mollusks are descendants of the snails that thrived in ancient Lake Idaho and other lakes of the Ice Age. When Lake Idaho drained, most of the mollusk species

Root structure of the desert sage.

perished. The few that survived are now threatened with extinction by pollutants and warm water temperatures.

While the middle Snake has been dramatically altered, stretches of free-flowing water between the 11 dams offer a glimpse of its former condition. Biologists consider the 8-mile stretch of free-flowing river between Lower Salmon Falls Dam and Bliss Dam as the best remnant of middle Snake ecology.

This stretch of river is lined with a unique deciduous riparian woodland dominated by hackberry and water birch trees. An additional 150 species of riparian herbs, grasses, perennials and annuals have been documented in this reach. Mammals include coyote, bobcats, mule deer, raccoon, beaver, mink, weasel, striped and spotted skunk, yellow-bellied marmots, porcupine, blacktail jackrabbit, river otter and various smaller rodents. A National Science Foundation-sponsored study of this river segment identified 83 bird species. Notable for their rarity are the white pelican, snowy egret, black-crowned night heron, ferruginous hawk, prairie falcon, osprey, Caspian tern and burrowing owl.

THE PLATEAU

While the river defines the plain, most of the area is an arid plateau. However, what most people see of this desert as they drive by on highways and roads is no more an accurate picture of Snake River Plain ecology than a clear-cut is of the Pacific rain forest. Much of the plain has been converted to irrigated farmland or is managed as rangeland for livestock, turning a complex desert ecosystem into crops, imported grasses and weeds.

But we can still see a remnant of the precivilization plain in remote places like the kipukas of the Great Rift. These islands of high ground surrounded by relatively recent lava flows (*kipuka* means "window" in Hawaiian) feature vegetation that is relatively lush, with native bluebunch wheatgrass, Sandburg grass, sagebrush, balsamroot, phlox, currant and other species creating a mosaic that supports a community of insects, reptiles, mammals and birds.

Domes of reddish-brown soil and rock chips mark colonies of harvester ants. Vegetation around the ant hills has been removed by the ants to allow more sun to reach the mound, which serves as a solar collector. Tall mounds of twigs mark the nests of formica ants notable for their ability to spray formic acid to deter predators.

Among boulders splotched with orange and yellow lichens rests the Great Basin rattlesnake, waiting to slither out at dusk in search of pygmy cottontails, voles and mice. Black raucous ravens wheel in the sky over the kipukas, and prairie falcons and red-tailed hawks occasionally soar out across the lavas. Meadowlarks sing their liquid notes from sagebrush branches, and mountain bluebirds dart in undulating flight. Along the ridges of the kipukas, the tops of desert parsley have been snipped off by mule deer. Surprisingly

ORD'S KANGAROO RAT

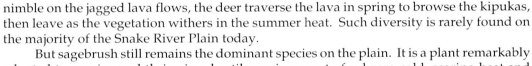

The Ord's kangaroo rat is a familiar sight as it races across desert roads at night. A tiny animal, it has adapted to arid and semiarid environments and does not need to drink water, since it is able to extract it from the vegetation and seeds it eats.

Kangaroo rats have fur-lined cheek pouches, a feature whose function is not understood. With long hind legs and weak, small front legs, they are called kangaroo rats because they run rapidly and erratically on their hind legs when scared by predators such as owls, coyote, or bobcats. Ord's kangaroo rats weigh between 1.5 and 2.5 ounces, are about 4 - 4.5 inches in length and have a long tail (up to 6 inches) with a tassle of hair at its tip. They are tan with a white belly. With their long whiskers, big eyes and long tail, they are an attractive animal.

Geographically, they are the most widely distributed of any of the kangaroo rats. This species is a granivore, living primarily on seeds produced by annual desert plants. Preferring sandy soils, they build their nests in burrows. Ord's kangaroo rats bear their young in late spring, and have litters of two to five. In some years they are able to produce more than one litter.

Kangaroo rats dig in desert sand to find buried seeds, and the small holes they leave as well as the trails made by their tails are common sights. Their home territories are up to several hundred square yards. In addition to the Ord's kangaroo rat, the Great Basin kangaroo rat also lives in southern Idaho. Although called kangaroo "rats," they are unrelated to European rats or to the native woodrat.

---

nimble on the jagged lava flows, the deer traverse the lava in spring to browse the kipukas, then leave as the vegetation withers in the summer heat. Such diversity is rarely found on the majority of the Snake River Plain today.

But sagebrush still remains the dominant species on the plain. It is a plant remarkably adapted to survive and thrive in a hostile environment of subzero cold, searing heat and frequent drought. Several species and subspecies of sagebrush can be found on the plain: Wyoming, mountain big, Great Basin, three-tip, low, black and silver sagebrush. Each species inhabits slightly different habitat created by varying soil depths and moisture. Early settlers learned to look for mountain big sagebrush when seeking potential farm–land; it grew in the deepest soils. Silver sage is tolerant of saturated conditions and indicates areas that occasionally flood.

Sagebrush features an hourglass-shaped root system, with shallow branching roots up top, and a tap root which drives deep into the soil before flaring out at its base. The tap root will grow as deep as the unconsolidated soil: 5 feet is typical for mountain big sage. This root structure allows sagebrush to tap not only more available water sources, but soil nutrients as well. The sagebrush is semievergreen, retaining a portion of its leaves throughout the winter. It is able to photosynthesize at very low temperatures, and thus will produce sugars on a sunny winter day. Sagebrush is also blessed with a sophisticated chemistry, exuding chemicals that deter most insects and make it similarly unpalatable to most mammals and birds.

Despite its fundamental place in the ecology of the plain, sagebrush has been the subject of ongoing eradication campaigns. In an attempt to turn much of the plain into livestock pasture, sagebrush has been poisoned with chemicals, burned and ripped from the ground by ship-anchor chains dragged behind pairs of bulldozers. In its place exotic grasses like crested wheatgrass, a Russian import, have been planted.

Even where sagebrush has not been deliberately removed, poor grazing practices have created conditions for wildfires, which burn out vast areas of sagebrush. The cycle starts with overgrazing on a sagebrush ecosystem, which removes much of the forbs and

Rattlesnake

bunch grasses growing between the sagebrush. The loss of these plants contributes to an overall drying of both the soil and remaining plants, since there is less protective covering on the ground. Next, cheatgrass, an aggressive exotic annual from Russia, invades the spaces where the native vegetation formerly grew. Cheatgrass grows lush and green early in the spring, but quickly turns dormant and dry. By midsummer a carpet of tinder dry cheatgrass is ignited by lightning storms. The fires burn the sagebrush, destroying its ability to reseed. The next spring the cheatgrass returns even thicker throughout the expanse of burned rangeland. Summer wildfires may again torch the cheatgrass, spreading flames to adjacent stands of sagebrush. Thus the cheatgrass creates ever-more territory for itself while eliminating all other forms of vegetation.

The pronghorn antelope, one of the swiftest mammals in North America, can reach speeds greater than 60 miles per hour.

The result of this cycle has been studied at the Birds of Prey Natural Area in the western Snake River Plain. As the sagebrush has been destroyed by fire the jackrabbit population has shrunk. Since jackrabbits are the primary prey of golden eagles, the eagles have not reproduced to their capacity.

Removing sagebrush from the plain is akin to removing trees from the forest. The wildlife and plants of each community are tied to these basic components. On the windy expanse of the plain, sagebrush also serves to hold blowing snow, allowing it to melt into the ground. But when the sagebrush is removed, so goes the snow, creating an even drier landscape.

Native grasses on the plain tend to be perennial bunch grasses, such as bluebunch wheatgrass. These grasses grow in clumps, densely occupying small spaces where they can send down longer roots to seek water. On the moister, higher elevation uplands of the Snake River Plain grows the juniper, a relatively short (20-foot-high) gnarled tree. Junipers are found in woodlands on the slopes of the Owyhee uplands and in small clusters on the Great Rift. The massive junipers of the Great Rift are hundreds of years old and grow in rocky crevasses virtually devoid of soil.

Much of the wildlife on the plain, notably sage grouse and pronghorn antelope, is singularly adapted to sagebrush. The pronghorn antelope are so adapted to the open plains that they do not know how to leap the fences erected across their range. They evolved in a landscape that had nothing higher than sagebrush. And they prefer the lower-growing dwarf and black sagebrush, avoiding the taller big sagebrush. But they can run like the wind, maintaining speeds of 40–50 miles per hour, and accelerating to over 60 miles per hour in short bursts.

Pronghorns are, in fact, not antelope at all, but the sole representative of the family Antilocapridae. They are unique in that they shed their horns each year, just as members of the deer family shed antlers. The pronghorn's horn consists of a bony core encased by a sheath of fused hair.

The coyote, a remarkably intelligent animal, hunts individually and in packs.

The pronghorn's coat is comprised of stiff, hollow hair that provides insulation against the winds that blow across the plains where it winters. The white rump patch hairs are twice the length of the rest of the coat hairs, and are controlled by specialized muscles. When alarmed, the pronghorn flares the rump patch hairs, doubling the size of the patch and sending an alarm signal to all nearby pronghorn. The pronghorn has exceptional eyesight and is always on the alert for danger, since it lives in open, exposed country.

Pronghorn populations on the plain have been reduced by nearly half since 1988 after they came in conflict with farming and ranching. The summer of 1988 was very dry, forcing the pronghorn to migrate early from the higher mountain basins of the Lost River and Lemhi ranges to their wintering grounds to the south. In the process the pronghorn grazed in irrigated farm fields. Farmers responded by persuading the Idaho Legislature to pass a law requiring the Idaho Department of Fish and Game to reimburse landowners for crop damage caused by wildlife. Faced with mounting depradation bills, the Department of Fish and Game adopted liberal hunting seasons to reduce population levels.

The construction of Interstate 15 also created problems for pronghorns, cutting off the herds east of Island Park from their wintering grounds. Following a series of mild winters, a herd may grow to 600–800 animals. But given a few hard winters, the population may plummet to less than 100.

Mule deer are the most abundant and widely distributed game animal on the plain. One of the most adapt–able of the game species, the mule deer has enlarged its range from the foot-hills and moved onto the plain. Presently there are about 100,000 animals that migrate back and forth from the foothills to the plain.

One animal that was eliminated from the plain and later returned is the California bighorn sheep, a more slen-der relative of the Rocky Mountain bighorn. At home among the canyons of the western Snake River Plain, the bighorns were destroyed by a combi-nation of market hunting and scabies, a disease introduced by domestic sheep. In the 1960s the Idaho Depart-ment of Fish and Game released Cali-

Bighorn sheep, top, thrive in the Owyhee uplands; mule deer, below, are found throughout the Snake River Plain.

fornia bighorns from eastern British Columbia into Little Jacks Creek canyon. Subsequent releases and a naturally expanding population have swelled the reintroduced herd to the upper and lower Owyhee River canyon and Deep Creek. This population has been used as a base for trapping and reintroduction in other areas of southern Idaho and in other states.

Predators on the plain today are limited to the prolific coyote, the badger, the bobcat and, in some areas, the cougar. Coyote do well because they are remarkably adaptable and intelligent, preying on everything from insects and berries to carrion and rabbits and occasionally deer and lambs. They hunt individually, listening for the movement of voles under the snow and then pouncing on the unsuspecting creatures. And they hunt in packs, running down young or weak deer and pronghorn, biting the neck for the kill.

Coyote were once poisoned wholesale with steer and horse carcasses laced with the poison 1080, which was banned because of its indiscriminate killing of other wildlife, such as eagles.

Badger

The burrowing badger, a member of the weasel family, lives wherever the soils are loose enough for digging and where its prey, the ground squirrel, lives.

Bobcats are solitary hunters, preying on rabbit and other small mammals and an occasional deer. Its larger cousin, the cougar, inhabits the canyons of the Owyhee, Bruneau and Jarbidge rivers, the higher country of the Owyhee uplands, the upper Snake River and the tributaries of the western plain. These large predators prey primarily on mule deer.

Gone from the plain are the gray wolf and the grizzly bear and the bison they once preyed upon. Wolves once roamed virtually all of North America, but an unrelenting campaign drove the wolf to extinction in nearly every state, including Idaho. Only in recent years has the wolf begun returning to Idaho — and here primarily in the dense and remote forests of the central Idaho wilderness areas. The grizzly bear too was eliminated by hunting and a loss of habitat and prey. Skulls of three grizzly bears were found in a lava cave near Craters of the Moon National Monument. Grizzly remains have been found in other lava tubes on the plain.

Bison were probably never as numerous on the plain as they were on the Great Plains of Montana, Wyoming, Kansas, Colorado and the Dakotas. Yet their numbers were sufficient that Indian sites throughout the plain contain their remains. The Baker Cave site on the Wapi Flow was used to butcher bison killed in winter. But, as in other states, hunting in the early 1800s virtually eliminated all bison from the plain.

Grizzly skull, from a lava cave near Craters of the Moon.

Prairie falcon with prey: rabbit and ground squirrel.

The Snake River Plain is justly famous for its raptors, with the Birds of Prey Natural Area the most notable example of prime habitat. In southwestern Idaho the Snake River has carved a deep, wide gorge through basalt, with cliffs providing innumerable nesting sites. Most raptors will nest in cliffs, including red-tailed hawks, ferruginous hawks and golden eagles. Prairie falcons will nest *only* in cliffs. Such nests, little more than a flat surface in a rock crevasse, are called scrapes.

The desert lands of the Snake have more prairie falcons than any place on earth, in part because of the abundance of nesting sites. But the plain also has a healthy population of Townsend ground squirrels, upon which the prairie falcons feed almost exclusively. The plateau on the northern side of the Snake River in the Birds of Prey Natural Area has deep, loose soils that are attractive to burrowing rodents. Prairie falcons hunt this plateau, flying downhill with their prey to feed their young. By midsummer the Townsend ground squirrels estivate — the summer counterpart to hibernation in which animals enter a torporlike state to pass through the dry, hot months. By then falcon nestlings have fledged and left the area.

Golden eagles also nest in the region, preying on jackrabbit and other small mammals. Eagles typically lay two eggs. However, the first chick to hatch has a head start on its younger sibling, and will usually kill its younger, smaller siblings, thus ensuring that it will receive all the food brought by the parents. Other raptors on the plain include red-tailed hawks, northern harriers, ferruginous hawks, Swainson's hawks and rough-legged hawks. While some raptor species nest on the plain and migrate south for winter, the rough-legged hawks nest in the Arctic and spend October through April hunting the white, windswept plain. Owls on the plain include barn, great horned, short-eared, long-eared, screech and burrowing owls that inhabit old badger or ground squirrel burrows and feed upon deer mice and insects.

Bald eagles and osprey, both of which prey upon fish, are found primarily at the margins of the Snake River Plain — along the Henry's Fork and South Fork of the Snake in the east and along the Payette and Boise river system in the west. These raptors require relatively shallow, clear water to spot fish from aloft. The uniformly deep channel of the Snake River does not provide that kind of opportunity.

Sage grouse noisily flap their wings during a mating dance.

If there is one animal most inextricably tied to the sagebrush desert, it is the sage grouse. The only bird to feed on the leaves of sagebrush, the sage grouse lacks a gizzard to digest hard foods such as seeds. Throughout the winter, sagebrush serves as the primary food of these large grouse.

Each spring, the grouse return to open areas within sagebrush stands — ancestral strutting grounds called leks. Here they dance at dawn to establish territory for mating rights with the assembled female grouse. During the short dance, the male sage grouse struts forward a few steps with head held high, his yellow eye comb expanded, white

Bear Paw Kipuka, an ancient caldera surrounded by younger lava flows.

The American bison, hunted to near extinction in eastern Idaho, still grazes the grassy highlands of Yellowstone National Park. Today the park is home to about 3,000 bison in three separate herds.

neck feathers erect, wings held out horizontally and tail feathers fanned. An air sac on the chest is inflated like a balloon and then immediately collapsed, creating two sharp snapping sounds. Males may square off on the border of territories. Standing side by side, but facing opposite directions, they flail at each other with their wings. Shortly after sunup the grouse disperse. The early morning gatherings continue until hierarchies are established and the females are impregnated. The males and females then separate, the females nesting alone beneath the sagebrush.

CONCLUSION

From the top of Old Juniper Kipuka on the southern end of the Great Rift, the Snake River Plain appears primordial, a volcanic landscape rolling on forever in distance and time. But that is a mirage. For millions of years the plain has undergone dramatic transformations, at times literally exploding. For millions of years the ecology of the plain has evolved along with these geologic and climatic disruptions. But today, people are the primary agents of change, and the alterations wrought by Western civilization have occurred over a mere 150 years.

Such swift and profound change has certainly damaged the ecological integrity of the plain: expanses of exotic cheatgrass overwhelming once diverse stands of bunch grasses and sagebrush; the great Snake drawn dry by diversion dams; cattle replacing herds of antelope and bison. But the plain will recover, long after we are gone, when a thousand years of our time is compressed into a layer of rock and sediment on the walls of the Snake River canyon.

Angered, the serpent tightened its coils. The pressure
became so great that stones began to melt. Liquid rock
flowed down the sides of the mountain. The huge
serpent, slow in its movements, was
roasted in the hot rock.

— *"Craters of the Moon"*
Shoshone legend

NATIVE TRADITIONS

NATIVE PEOPLES first encountered the Snake River canyons perhaps
12,000 to 15,000 years ago. Since that time they have successfully adapted to
the cold, relatively inhospitable environment of the Ice Age — the most recent geological epoch — and to the contemporary temperate climate. These
earliest Idaho peoples may have witnessed events unparalleled in recent
history. It is possible, for example, that they witnessed the awesome

BY MARK PLEW

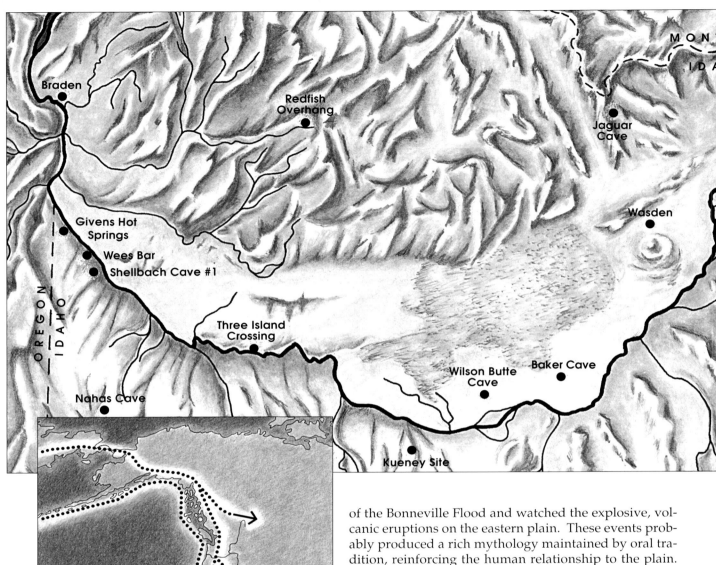

Pleistocene migration routes; previous page, petroglyphs from Birch Creek; detail, Shoshone pot.

of the Bonneville Flood and watched the explosive, volcanic eruptions on the eastern plain. These events probably produced a rich mythology maintained by oral tradition, reinforcing the human relationship to the plain. Although this oral tradition is not preserved in the archaeological record, much is known of the lifeways of these early peoples. From the remnants of their tools and a range of sites documenting their varied use and modification of the landscape, we see a highly successful adaptation that continues in the traditions and lifeways of contemporary American Indians.

The ancient story of humans and their relationship to the land is told in several archaeological sites that are scattered across the Snake River Plain. Some sites consist of a few isolated artifacts or the fragmented remains of a camp or village. There are also quarries and places where people made tools. Taken together, these sites are pieces of the complex puzzle that is the cultural history of the Snake River Plain.

In southern Idaho and throughout much of the western United States archaeologists divide time into two major cultural periods. The Paleo-Indian period (7,000–15,000 years ago) dates from the late Ice Age, or Pleistocene Epoch. Sometimes called the big-game hunting tradition because of its emphasis upon the use of large herd animals, the Paleo-Indian period was followed by an Archaic period (7,000 to historical times) in which humans made the transition from a hunting culture to one that diversified its use of an increasingly arid environment.

Major archaeological sites. Burials, quarries, camps, villages and other prehistoric sites yield fragments of tools and other artifacts — a rich material record of ancient life on the plain.

Mortar and pestle with Shoshone pot (replica). Stones and pots such as these were used to grind and store various seeds and bulbs.

EARLY MIGRATIONS

Traditionally, archaeologists have placed the migration of people into North America at no more than 20,000 years ago. Under the most common theory, hunters from Siberia traveled across the Bering Strait land bridge, created during the Pleistocene Era when glacial ice lowered the water levels of the world's oceans. The bridge between Siberia and Alaska, sometimes called "Beringia," was nearly 1,000 miles wide. The first ancestors of modern American Indians were probably hunters of Eurasian origin who brought with them material culture developed in the Old World. In addition to early peoples, a number of animal species, including elephant, deer, elk and moose, migrated across the strait. As these species moved east, the camel and horse, which evolved in the Americas, moved across the land bridge to Eurasia.

While this theory is widely embraced by scholars, increasing archaeological evidence suggests alternative and earlier migrations into the Americas. Sites excavated in Mexico and South America suggest a far earlier human presence. It seems likely that some people from Eurasia may have traveled by boat down the North American coast. But as is the case with much of the Snake River Plain, the rapidly changing geology of the Pacific Coast may have obliterated any evidence left by these early nomads.

Although it is difficult to determine when people arrived in the Americas, scholars say with certainty that by 12,000 years ago humans lived on the Snake River Plain.

Rising oceans and the changing geology of the rugged North American coastline may have obliterated any evidence of a possible coastal migration from Asia.

THE PALEO-INDIAN TRADITION

The large glacial ice masses had far-reaching effects on the climates of North America. Areas near the masses were cold and moist, with a limited growing season, much like modern Arctic tundra conditions. Farther from the ice sheets, the environment was more hospitable. In some areas, including the Snake River Plain, large coniferous forests grew where there is now desert steppe country.

Little is known about the earliest inhabitants on the plain. The oldest artifacts from the Paleo-Indian period are simple modified bone and stone tools. Because so few artifacts have been found from this ancient period, the lifestyle and activities of these people remain unknown. Archaeologists, nevertheless, have placed them in three cultural periods — the Clovis, Folsom and Plano — based on the tools they made and the places where their artifacts have been discovered. Each of these groupings represents minor changes in technology.

The earliest evidence of human occupation on the Snake River Plain comes from a lava blister near Dietrich known as Wilson Butte Cave. Excavated in 1959–1960 by Ruth Gruhn as a joint Idaho State College–Harvard University museum expedition, the cave deposits suggest almost constant — but irregular — use during the past 15,000 years. In fact, some of the earliest evidence of human presence in North America comes from Wilson Butte Cave. The lowest level, or stratum, dated at 14,500 years old contained retouched flakes, a basalt knife and a bone fragment with cut marks. Archaeologists also discovered the remains of extinct camel, horse and sloth, which suggest somewhat cooler and moister conditions during the period 15,000–6,850 years ago.

The Clovis period, which dates between 12,000–11,000 years ago, was an era of big-game hunting. It has been assumed that the Clovis people hunted with what has become known as the Clovis point, presumably mounted on a spear. Perhaps. But it is questionable whether huge animals such as mammoths could have been seriously hunted by spear-wielding humans. It is more likely that the mammoths were taken only occasionally and opportunistically, such as if one of the giant elephants became trapped in a swamp.

Clovis peoples lived on the plain during a period of relative cold and marginal vegetation. They were hunters who may have lived in caves or temporary rock shelters and who sometimes pursued such large and now absent animals as the camel and bison.

Archaeologists have documented the Clovis period at several sites across southern Idaho, with materials reported as far east as the Portneuf River. The most significant is the Simon site near Fairfield, where archaeologists discovered a cache of Clovis points, bifaces and several partially worked slabs of smoky quartz. Because the cache is isolated and remnants of red ochre covered the points, the site may have been a place used for ritual or ceremonial activities.

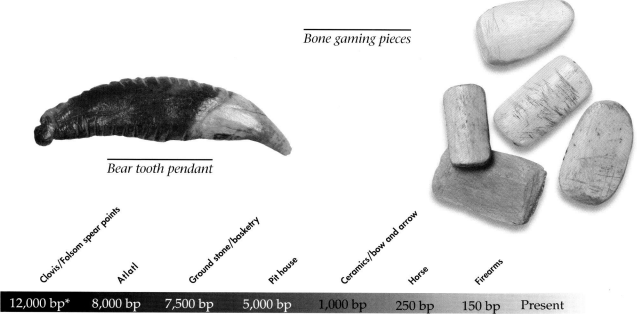

Bone gaming pieces

Bear tooth pendant

Clovis/Folsom spear points	Atlatl	Ground stone/basketry	Pit house	Ceramics/bow and arrow	Horse	Firearms	
12,000 bp*	8,000 bp	7,500 bp	5,000 bp	1,000 bp	250 bp	150 bp	Present

*Years before present (1994)

RUTH GRUHN
(1935–)

Ruth Gruhn began her career in Idaho archaeology as a graduate student at Radcliffe College. Her most significant contribution to Idaho archaeology was her excavation of Wilson Butte Cave, which documented the first major cultural sequence in the state. Importantly, Gruhn documented the association of human artifacts with extinct camel and horse remains dating from 14,500 years ago to relatively recent Shoshonean occupations. Gruhn and her husband, Alan Bryan, spent three decades conducting pioneering research throughout South America. They then returned to Idaho to reinvestigate Wilson Butte Cave. Their second excavation indicated that human activity in the cave may have been more recent than was originally thought, dating from about 10,500 years ago.

Gruhn at Wilson Butte Cave in 1959 (above) and 1989.

Killing a mammoth, a drawing by illustrator and naturalist Charles R. Knight. Knight's illustration, first drawn for McClure's magazine in 1897, reflects the 19th-century assumption that Clovis hunters commonly preyed on huge creatures. Today many scholars believe that large mammals such as the mammoth could be killed only when sick, crippled or mired in a marsh.

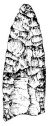

Large channel flakes are removed from Clovis points in a process called "fluting."

Wilson Butte Cave and Jaguar Cave, located in the Lemhi Mountains of eastern Idaho, are also significant Clovis sites, although neither contained Clovis points. Jaguar Cave held the butchered remains of 268 large mountain sheep, killed between 11,540 and 10,270 years ago. The remains of domesticated dogs — at 9,400 years old among the earliest in the world — were also discovered at Jaguar Cave. Mountain sheep are not normally associated with the Clovis people, but the dog remains suggest a long-term use of dogs in the hunting of mountain sheep, as depicted in the historic literature for mountain Shoshone.

In contrast to the Clovis, the Folsom period, 10,600–11,000 years ago, is better documented in southern Idaho. The Folsom peoples used smaller fluted projectiles, primarily to hunt the now extinct bison, *Bison antiquus*, which became more prevalent at the end of the Pleistocene.

Folsom points are smaller than those from the Clovis period, and fluting extends nearly the length of the projectile.

The Folsom way of life was probably similar to that of the earlier Clovis peoples — a nomadic existence based on hunting of animals. But it also included an increased use of plant foods. Hunting was probably done with spears after game had been driven into compounds or canyons. Like the Clovis, the Folsom peoples probably lived in temporary shelters.

Archaeologists document the Folsom period by many surface finds. At the Wasden site, also called Owl Cave, near Idaho Falls, a lava tube contained the remains of camel, bison and mastadon alongside four Folsom points dated between 10,920–12,850 years ago. The Wasden site, one of the earliest Folsom sites in North America, may have been used as a natural corral for bison that were killed and butchered there. The site is unusual in that mastadons are not normally associated with the Folsom people.

Plano points are large and not fluted.

The Plano period, 7,800–10,600 years ago, is well documented from surface finds and excavations. Many, in contrast to the Clovis and Folsom periods, appear to include habitation sites in addition to kill or hunting sites. The Plano peoples used caves and rock shelters as places to live, which suggests greater permanency than the earlier periods. More specialized tools and artistic ornaments also indicate that these people may have been involved in more diverse economic pursuits than simple hunting.

Excavations have uncovered Plano-type projectiles that were used to hunt bison and sheep. Plano materials were recovered from the upper levels of the Wasden site in association with *Bison antiquus* remains. The skeletons of some 70 individual animals suggest two separate kills — at the beginning and end of the calving season.

At Wilson Butte Cave some 120 miles west of Wasden, archaeologists have found a number of Plano-type projectiles along with milling stones. The discovery of a mano, a grinding stone used to process seeds and other wild plant foods, suggests that the Plano peoples began to diversify their economy by using plant foods in addition to the big game they had traditionally hunted.

Atlatl (replica) with weight.

ARCHAIC TRADITION

By the end of the Paleo-Indian period some 7,000 years ago, the glaciers were retreating and the climate was becoming warmer and dryer. With this change in climate came an apparent change in the hunting and gathering patterns of the people. Unable to adapt to differences in vegetation, the elephant, camel and horse had become extinct in North America. Smaller animals better adapted to a dry environment flourished, and the people responded by developing tools to hunt these animals, as well as to forage for plants. This adaptation to a more diverse resource base marked the beginning of the Archaic period.

Known also in the West as the Desert Culture, the Archaic period saw the development of tools not found in the preceding Paleo-Indian culture. Included were grinding or milling stones used to process seeds, woven sandals, moccasins, twisted cords, basketry, and wooden and bone tools including punches, drills, wrenches and digging sticks.

The most significant weapon was the spear-throwing atlatl, a small wooden shaft with a spur on one end, fitted with fingerloops and weights to increase balance and thrust. The end of the spear was placed against the spur shaft, and with the hand on the other end the spear was launched, with more force and accuracy than was possible throwing by hand. The atlatl increased the ability of Archaic hunters to stalk animals such as the antelope.

The early Archaic period featured large corner and side-notched points used with the atlatl.

Decorative, recreational and ceremonial items became more common during the Archaic period. Gaming pieces, ornaments, molded clay objects and figurines have been found in a number of Archaic sites. Olivella shells from California and seashells from the Northwest coast suggest widespread trading among native peoples.

The Archaic period probably represents the beginning of seasonal migration patterns known as "transhumance." For example, groups might spend winters in the stream valleys hunting and fishing and return in the spring to higher meadows where green plants could be harvested. In late summer, they might move to small canyons to collect and process wild fruits and berries, eventually returning to their winter camps to begin the cycle again.

This pattern was characteristic of the historic Shoshone Indians who inhabited much of southern Idaho in late prehistoric and early historic times. The Shoshone scheduled their

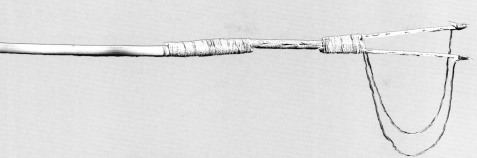

Harpoon (replica). Detachable tips were wrapped with cordage then waterproofed with a mixture of pitch and ash.

FISHING

In the traditional view of middle Snake River fishing, groups descended on certain stretches of the river to harvest and store salmon for winter. This view is based on the ethnographic record of the 1930s and earlier historic records of explorers, trappers, soldiers and migrants. The records for southern Idaho are sparse in their depiction of the lifeways of the aboriginal population, however. The few existing accounts reflect the "memory culture" of individuals who did not live the aboriginal lifestyle of the pre-reservation period. Some archaeologists believe that salmon fishing was not the most optimal economic strategy. Although the salmon were often abundant, other kinds of food may have been less difficult to process and store. Abundance alone does not determine usage. Thus, fishing reflected the diversity of aboriginal lifeways in which a variety of factors determined the sources of food.

Bone hook and plant fiber cordage.

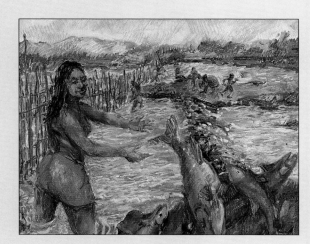

Collecting fish from a weir.

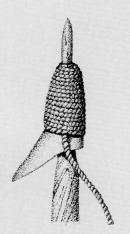

Fishhook

PREPARING THE FOOD

Camas Roots

In the spring and summer when the camas flowered, women would dig up the roots, then place them in an earthen oven. This oven was a pit lined with hot rocks and filled with alternating layers of camas roots and grass, then covered with soil. The women built a fire over the pit and cooked the roots for several days.

Chokecherries

Indian women valued the chokecherry bush for two things: the stems, brewed into tea, and the bright red berries, which were mashed into cakes. After mashing fibrous berries with a flat, boardlike tool and a hand-held grinding stone, the women would make reddish patties that were then sun-dried.

Seeds

Beaters and winnowing trays were used to collect a range of seeds. These were commonly ground, mixed with other items and formed into cakes left to dry in the sun. Seeds were sometimes parched for storage.

Brush hut, an early housing style; below, one of eight prehistoric foundations discovered near Marsing.

movements according to the availability of specific resources. They depended heavily upon salmon and other fish, particularly when wintering in the Snake River canyon, where they stored root caches and dried meats. In the winter, they hunted sage hen, snowshoe rabbit, grouse and deer. Fish and sufficient supplies of firewood were available year-round near the Snake River. In spring, green plants and roots flourished. Two spring salmon runs provided a principal food source for the aboriginal inhabitants of the western Snake River canyon. Fishing for salmon was a cooperative undertaking, generally within family units and particularly where small villages were located in fishing streams near the canyon. Fish were speared or taken with nets and hooks. Sometimes several families built a dam or weir and in turn distributed the catch among themselves, giving a share to a director who coordinated their efforts.

It is known that the Snake River Shoshone traveled in July to Camas Prairie, some 35–50 miles north of the canyon to collect camas and other root crops as well as to hunt small mammals, like the gray ground squirrel. Many of the root crops were dried, returned to the canyon and stored in cache pits for winter use. At the end of summer the Shoshone returned to the Snake River to harvest fish from the fall salmon run. Much of the catch was dried and smoked on wooden racks and stored for winter use. Sometimes they cooked and mixed salmon oil and formed it into a pemmican and stored it in salmon skin bags. In the fall some Indians traveled to the surrounding hills and small drainages to collect seeds and chokecherries. Chokecherries, like a variety of other foods, were ground or pounded on a metate or mortar, dried into small cakes, and stored in cache pits with other foods. Some seeds were collected and parched to prevent germination during storage. The Shoshone Indians, as well as other historic groups on the plain, did not grow corn.

THE EARLY AND MIDDLE ARCHAIC

The Early Archaic period (5,000–7,800 years ago) marks the transition from the Plano to the Archaic, a time when subsistence and material culture changed substantially.

The Snake River Plain is rich in Early Archaic sites including Wasden, Wilson Butte Cave and Birch Creek northwest of Idaho Falls. In western Idaho, archaeologists have found archaic materials from the Owyhee Uplands, the Castle and Brown's creek areas, Reynold's Creek drainage, areas along the Snake River near Marsing and in eastern Owyhee County. Northwest of Boise, Early Archaic materials have been found in the Payette River drainage and in the Weiser area.

Several finds indicate people during the period hunted modern animals such as bison, deer, elk and mountain sheep. Evidence of plant use is also relatively common. Some of the earliest evidence comes from Wilson Butte Cave and the Owyhee Uplands at Nahas Cave, where milling stones date to approximately 6,000 years ago.

During the Middle Archaic period (1,000–5,000 years ago) people used groundstone implements more extensively, and there is evidence of greater diversity in settlement-subsistence patterns. Sites have been discovered in a range of river, foothill and upland

Sandal, made of woven sagebrush.

Juniper and obsidian knife (replica).

settings. There is some archaeological evidence that indicates that individual sites were used for specific purposes, such as fishing, mussel collecting, seed collecting and grinding or bison hunting.

The first houses on the plain date from the Middle Archaic period. In 1982 archaeologists found eight house structures containing deer and mussel remains at Givens Hot Springs near Marsing. The houses were of two types. The first houses were a series of steep-walled, semisubterranean structures about 23 feet in diameter and 1–3 feet deep. The second and more recent type of houses were also steep-walled, semisubterranean structures 12–21 feet in diameter and about 2 feet deep. Saucer-shaped, they show evidence of hearths and roof supports. One burned house had rafters aligned in a circular fashion around the roof supports and was thatched with heavy grass. These "wickiup" type structures were common in the Late Archaic period, about 1,000 years ago. Both house forms have also been found in the Great Basin and Columbia Plateau.

In western Idaho, the period is best represented at Nahas Cave, a hunting site in the Owyhee Uplands where Early, Middle and Late Archaic materials have been found along with the remains of suckers and steelhead trout. One mile from the Nahas Cave at the Deep Creek Rockshelter, archaeologists have found evidence of mussel collecting. Middle Archaic peoples also collected mussels at Rock Creek and Kueney in the South Hills country near Twin Falls.

EARL SWANSON
(1933–1975)

A pioneer in Idaho archaeology, Earl H. Swanson Jr. arrived at the Idaho State College Museum in 1957. He began an ambitious archaeological survey and excavation program that marks the beginnings of systematic archaeology in the state. Swanson brought together a team of experienced field archaeologists that included B. Robert Butler, Alan Bryan and Donald Tuohy. Swanson also earned an international reputation for geoarchaeology. His Birch Creek project in northeastern Idaho was noted for the use of geologic data in interpreting archaeological discoveries.

Swanson at Blue Lakes dig and with student.

SEASONAL MIGRATIONS

The Snake River peoples followed a seasonal or "transhumant" pattern of migration. In winter they might hunt and fish in stream valleys. In spring they might migrate to higher meadows to gather plants. They might spend late summer in small canyons collecting fruit and berries, returning again to their camps in the lower valleys at the onset of winter. This map illustrates how Archaic peoples moved from the canyons to foot-hills and meadow, exploiting seasonal resources.

Brush tepee, a summer shelter.

○ Winter Camp
△ Spring / Summer Processing Camp
□ Fall Hunting Camp

DON CRABTREE
(1912-1980)

Imagine an obsidian spear, knife or arrowhead sharper than a surgeon's scalpel. Don Crabtree, a self-taught toolmaker, rediscovered the lost art of crafting these razor-sharp points, turning a boyhood hobby into scientific expertise.

Born in Heyburn and raised in Salmon, Crabtree first took an interest in stone tools after a neighbor offered him an arrowhead as payment for running an errand. "I was intrigued," Crabtree recalled. "It became a challenge to duplicate them." Chipping at natural glass with a pointed hammer, Crab-tree replicated a wide variety of spear-points and blades, the deadly weapons once used on the mammoths, musk-oxen and bison that roamed the Snake River Plain. "Some aboriginal tools are really pieces of beauty," Crabtree explained. "They had a lot of Michelangelos in those days."

Pragmatic and largely self-educated, Crabtree was a University of Idaho research associate who worked closely with Idaho State University and lectured widely in Idaho, Washington, Canada, Central America, England and France. Writer, teacher, curator, craftsman, a giant in the emerging field of lithic technology — Crabtree remained, above all, a dedicated experimenter. During an experimental lung operation in 1976, Dr. Bruce Buck, a Twin Falls surgeon, cut into Crabtree's chest with one of his patient's tools, a fragile obsidian blade. Years later the toolmaker from Salmon unbuttoned his shirt for a *Statesman* reporter and dared her to find the scar.

"He could have taught ancient man a thing or two about toolmaking," said Francois Bordes, a French archae-ologist. Crabtree, however, was modest about his achieve-ments: "They [the ancient craftsmen] were doing things with this rock or volcanic glass that nobody knows how to do today."

Crabtree demonstrates flintknapping techniques.

At Bachman Cave in western Owyhee County, there was evidence of extensive ground stone use during the Middle Archaic. Further evidence of economic diversity includes the partially butchered remains of an immature bison, dating from 4,000 to 5,000 years ago, in the Snake River canyon near Bliss. Sites in the Payette River drainage date from 5,000 to 10,000 years ago.

Burial grounds located near Weiser suggest a growing social complexity in the Middle Archaic society. The sites feature burials and cremations, with exotic grave goods buried with the individuals. The dead were often buried with large stone blades called "turkey tail" points. Other grave goods included pipes, hematite crystals, marine shells, dog remains and extensive use of red ochre. The inclusion of exotic, sometimes "imported" materials in burials and selection of sandy knolls and terraces as burial locales indicate preferential treatment of the dead.

The western Idaho Archaic burial complex is especially important because it docu-ments a trend toward greater trade networks and suggests that Middle Archaic popula-tions were becoming more socially complex.

THE LATE ARCHAIC

During the Late Archaic era (340–1,000 years ago) life on the Snake River changed as new technologies were developed and introduced. The bow and arrow and ceramic technology were introduced during this period, and the economy became more diversified as people made more extensive use of the Snake River.

Beginning about 1,000 years ago, the use of small corner and side-notched projectiles referred to as Desert Side-Notched and Rose Spring–Eastgate points became increasingly common. These points are believed to represent the widespread replacement of the atlatl by the bow and arrow and presumably mark a shift toward the hunting of smaller mammals.

An additional, important contribution from the Late Archaic period is the introduction of pottery. Though evidence of fired clay technology dates as early as 6,000 years ago in the Owyhee Uplands, pottery does not occur in archaeological contexts until approximately 1,000 years ago. The common pottery type found throughout the Snake River Plain is a rather poorly made and largely undecorated, flat-bottomed vessel resembling a flower pot that is referred to as Shoshone Ware. This pottery or a knowledge of pottery making may have been brought to southern Idaho by Shoshone peoples migrating from the southwestern Great Basin.

The recent discovery of additional pottery-bearing sites and the reevaluation of ceramic collections have led to considerable debate regarding the origins, variation and cultural affiliations of the ceramics of the Snake River Plain. There may be some relationship between Idaho Shoshone Ware and Fremont pottery associated with the relatively sedentary horticulturists of northern Utah. Northward-expanding Shoshones may have carried with them Fremont-type potteries or more probably a knowledge of Fremont ceramic technology and styles. There is some evidence from Wilson Butte that the Fremont people may have occupied a portion of the eastern Snake River Plain. Fremont basketry also is present at a number of sites in the eastern plain.

By the Late Archaic, aboriginal populations had expanded and a number of new settlement-subsistence regimes were well established. Indeed, it appears from the archaeological and ethnographic records that some groups focused more on single resources. Such may have been the case with salmon fishing on the middle Snake River. Archaeolo-

In the Late Archaic, small triangular side-notched points indicate increased use of the bow and arrow.

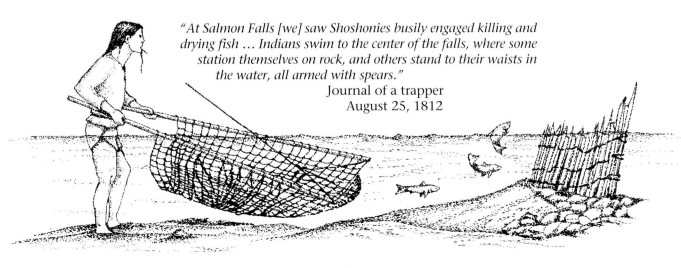

"At Salmon Falls [we] saw Shoshonies busily engaged killing and drying fish … Indians swim to the center of the falls, where some station themselves on rock, and others stand to their waists in the water, all armed with spears."
Journal of a trapper
August 25, 1812

Late Archaic peoples adapted to an increasingly hot and arid climate with brush shelters, woven clothing and baskets, and new kinds of weapons like harpoons with detachable tips and the bow and arrow.

Decorative clay figurines

gists have documented the earliest use of salmon at approximately 7,000 years ago at Bernard Rockshelter in Hells Canyon. In the Owyhee Uplands, steelhead remains have been dated at 3,000 years ago at Nahas Cave.

Along the Snake River there is greater evidence of fishing during the Late Archaic. At Schellbach Cave No. 1, south of Boise, archaeologists have found chinook remains and fishing gear, including net sinkers, rope and fishhooks. Two fish traps have been found in the Snake River canyon near Twin Falls, as well as a number of aboriginal fish weirs between Shoshone Falls and the Owyhee River in Oregon.

Recent excavations at Three Island Crossing near Glenns Ferry have recovered more than 19,000 fish remains associated with a house structure and storage pits, a locality described in the historic literature. The use of salmon as a major resource probably resulted in some Late Archaic populations becoming more seasonably sedentary. Evidence of this comes with the discovery of Late Archaic houses at Hagerman Valley, Big Foot Bar, Indian Cove and, as noted, Three Island Crossing.

The Late Archaic period was also rich in petroglyphs and pictographs, rock art that gives us a glimpse into the culture of these ancient peoples.

Though presumably present in the Early and Middle Archaic periods, rock art is notably more common in the Late Archaic. Petroglyphs — figures pecked or pounded into boulders or rock walls — are perhaps most common on the western plain, with examples along the Snake River at Wees Bar, in the Owyhee Uplands and in the Bennett Hills. Much of the petroglyphic rock art is typical of the Shoshone styles found in California and Nevada.

Pictographs — figures painted using a variety of natural pigments — are relatively more common in the eastern plain and depict a greater variety of natural- istic and historic period motifs. Rock art probably served a number of purposes, from magico-religious functions to the marking of hunting trails.

Hunting facilities — rock alignments formed by stacking stones into circles, walls and rimrock enclosures — are also common on the Snake River Plain. In the Owyhee Uplands, rimrock enclosures were used as corralling devices. Bison jumps have been reported near Challis in eastern Idaho and in eastern Owyhee county, although the latter most probably were used to hunt deer and antelope.

How long have the Shoshone lived on the plain? While some archaeologists believe that the Shoshone peoples have inhabited southern Idaho for thousands of years, others, noting the similarity between Snake River dialects and desert languages to the south, say Shoshone peoples arrived about 700 years ago. The Snake River Shoshone consisted of small bands of hunters and gatherers whose nomadic lifestyle and use of diverse resources scattered them about the plain.

Breathtaking cascades such as Twin Falls were important spearfishing sites. Englishman Edmond Green, a commercial artist, sketched this native spearfisher at the base of the falls on a trip to Idaho in 1880; below, Boise State University archaeological field school at another fishing site on the Snake River near Three Island Crossing, 1987.

EQUESTRIAN PERIOD

While Native Americans had been slowly adapting to environmental and technological changes for 15,000 years, the influence of white Euro-Americans precipitated sudden and rapid changes.

The first influences were created long before Indians in Idaho ever encountered Europeans. By 1750 A.D., the beginning of the Equestrian period, many of the Shoshone groups in southern Idaho had acquired horses, which were introduced by the Spanish in the Southwest and spread north through trade with the Utes. The horse vastly changed the way of life for those Indians who adopted the animal. With expanded territories, these peoples used new resources and developed wider trading networks. With horses, they could travel greater distances in search of food, giving them a competitive advantage over unmounted Indians. This may have threatened the monopoly of some resources by resident bands and, in other instances, encouraged a reliance on new resources, such as the salmon. The arrival of the horse had a great impact on some groups. The Bannock, for example, are Northern Paiute from western Idaho who moved to eastern Idaho after acquiring the horse. They hunted bison on the fringes of the Great Plains. Such interaction between Idaho Indians and peoples on the Great Plains resulted in the adoption of many items of material culture, such as saddles, tipis and clothing.

In general, populations increased during the Equestrian period, though it is not clear if this was directly related to the horse. So-called "villages" noted by explorers and later described by anthropologists may represent nothing more than communal seasonal camps for hunting, fishing or food gathering. Historic population estimates for western Idaho suggest no more than one person per 15 square miles. Though native groups congregated in some areas to seasonally use certain resources, the view that many peoples commonly found themselves on the edge of starvation may not be fully warranted. Resources were plentiful and the groups who relied on them were very small.

During the Equestrian period native peoples began to use Euro-American materials. Archaeologists have uncovered metal items at sites in the Birch Creek Valley and along the Snake River. A brass projectile was found at Three Island Crossing, and copper artifacts were discovered at Big Bar in Hells Canyon. At the Rattlesnake Canyon site near Mountain Home, copper fragments were recovered in association with cremated remains. Eventually, most Indians quit making and using arrows altogether in favor of newly acquired rifles.

Many groups on the western plain did not acquire horses and continued to practice earlier lifeways but the influx of explorers, trappers, traders, settlers and, eventually, the military ended the aboriginal way of life for all but the most remote groups. Living and traveling in small, dispersed bands, the Idaho Indians could not take a unified stand against the Euro-Americans. However, that isolation and independence may have saved them from the worst affects of smallpox and measle epidemics that devastated larger, more settled tribes, such as the Mandans and Pawnee of the Great Plains. But the environmental impact created by the invading white culture may have caused far greater havoc. By the middle 1800s, whites were clearing the river bottoms of trees for firewood and lumber. Huge herds of horses and cattle were overgrazing the grasslands in some areas of the plain.

Initially, such actions would have reduced the populations of deer and bison that comprised the main diet of the Indians. But the damage certainly would have resulted in erosion and the silting of streams and rivers, as well as a change in the plant life of the

Warrior on horseback, detail from an 1871 George Catlin painting; below, horse petroglyph near Gooding.

THE HORSE

Introduced to North America by the Spanish conquistadors, the horse reached the Snake River Shoshone in the mid-1700s, radically changing their lives. One authority believes that the Shoshone and Bannock peoples had more than one horse per capita by the time of Lewis and Clark, but some tribes had more horses than others. The Lemhi Shoshone selectively bred horses while, to the west, the Boise and Bruneau bands continued to forage on foot.

By 1800 the mounted Shoshone ranged widely, hunting the buffalo to near extinction in southeastern Idaho and crossing the Rockies to chase buffalo on the Wyoming plains. Contact with equestrian buffalo-hunting Great Plains Indians brought new technologies and fashions: tipis, tailored clothing with ornamental bead work, buffalo robes, feathered headdresses,

Woman's saddle

food preservation techniques, leather saddles, horseshoes and the horse-drawn sled with netting suspended between two poles, a "travois."

Horse culture changed Shoshone politics as influential chiefs took command of the buffalo hunts. The horse also revolutionized trade. By the early 1800s the highly mobile Lemhi Shoshone had made contact with the natives of the Glenns Ferry area, trading buffalo robes for fish. Each summer during salmon season the Boise Shoshone returned to the Weiser area for a trade festival. Paiutes rode in from Oregon and Nevada. There also were Nez Perce, Crow, Cayuse, and Umatilla traders. Some scholars believe these Indian festivals were the precursors of the trapper rendezvous of the fur-trade era.

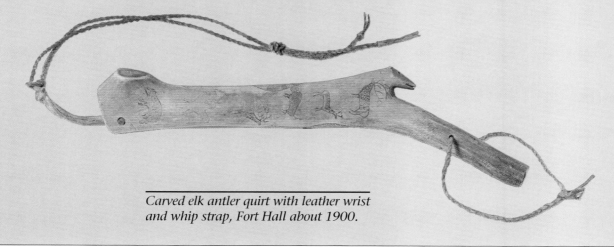

Carved elk antler quirt with leather wrist and whip strap, Fort Hall about 1900.

Chief Little Soldier,
Shoshone, about 1870.

plain. The Indians may have been adapting to these changes when the first whites began recording their observations of the native people. Thus, these first ethnographic records may represent an already drastically changed life from that of Indians just 100 years earlier.

By the end of the 19th century, the Indians of the Snake River were being forced onto reservations, where the U.S. Bureau of Indian Affairs tried to turn hunters and gatherers into farmers and herders. After 15,000 years of adapting to evolution on the Snake River Plain, the Indians were forced to make changes that were neither natural nor of their choice.

Wild cane and rosewood
arrows fitted with wild turkey
and goose feathers; mountain
mahogany bow (replicas).

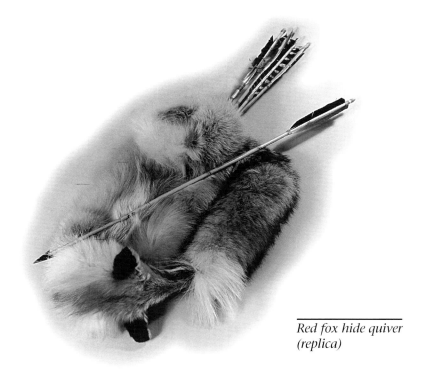

Red fox hide quiver
(replica)

Between Apple Trees.

It was a relief to see the distance widening between
us and those volcanic strata. It was a desolate, dismal
scenery… Not a shrub, bird, nor insect seemed
to live near it. Great must have been the relief of the
volcano, powerful the emetic, that poured forth
such a mass of black vomit.

— *Julius Merril*

Trail journal, 1864

CONFRONTING THE DESERT

TRAVELERS PASSING THROUGH southern Idaho are amazed
by the varied terrain. High mountain valleys gradually give way to
irrigated fields of potatoes and grain. Where the Snake River flows into
south-central Idaho, there is a vivid appearance of agricultural prosperity. Field after field receives periodic dousing from mobile sprinkler

BY F. ROSS PETERSON

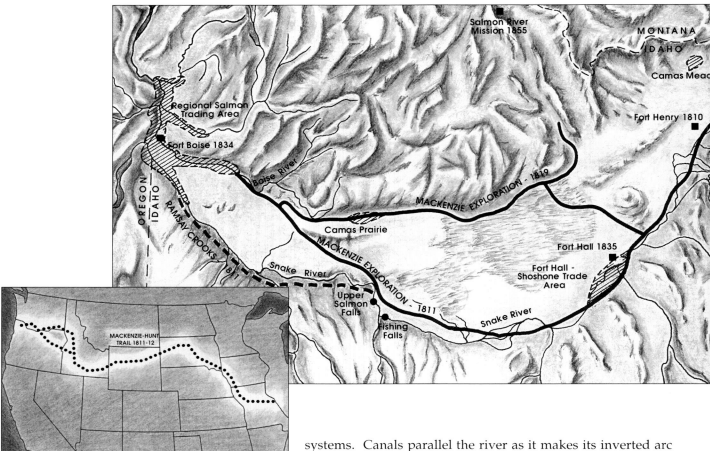

Labels on map: Salmon River Mission 1855; MONTANA; IDAHO; Camas Mead; Regional Salmon Trading Area; Fort Henry 1810; Fort Boise 1834; Boise River; MACKENZIE EXPLORATION - 1819; OREGON; IDAHO; RAMSAY CROOKS; Camas Prairie; Fort Hall 1835; MACKENZIE EXPLORATION - 1811; Snake River; Fort Hall - Shoshone Trade Area; Upper Salmon Falls; Fishing Falls; Snake River; MACKENZIE-HUNT TRAIL 1811-12

*MacKenzie-Hunt trail,
1811-1812; previous
page, Twin Falls
potato pickers and
diggers, about 1910
(colorized); detail,
state seal from Vardis
Fisher's Idaho
guidebook, 1937.*

systems. Canals parallel the river as it makes its inverted arc through Idaho. Many farmers still flood their land by opening a headgate and allowing the water to penetrate the rows of dry beans, wheat, corn, alfalfa or potatoes.

This fertile land bears little resemblance to the sagebrush plateau that dominated southern Idaho before settlement and irrigation. Even today, large tracts of sagebrush and bunchgrass remain. Where there is no water, there is little life.

While seminomadic tribes found ways to exploit the desert, the desolation repelled the first waves of Euro-American civilization. Early explorers and trappers, chiefly concerned with survival, dismissed the barren plateau. The first homesteaders preferred the green valleys of Oregon and California. Not until the discovery of gold in the 1860s did farmers regard the desert as land to be valued for growing crops to feed thousands of hungry miners. And not until the Mormons settled the plain did Idahoans recognize the enormous potential of ditch irrigation.

As ranching and farming took root, demand for water increased. Water developers and land speculators brought in federal money for some of the nation's most massive dams. This development, for the most part, is of 20th-century vintage, and it is one of the West's most remarkable stories of human adaptation and interaction with a once barren land.

FIRST ENCOUNTERS

Culturally and geographically, there are at least two Idahos — a northern "plateau" region, where Lewis and Clark met the Nez Perce, and a southern "plains" region, bounded roughly by the Salmon River, Hells Canyon and the Wyoming Rockies. The largest group

Fur-trade frontier, 1810–1835. Donald MacKenzie and Wilson Price Hunt led the 1811 Snake River expedition that pioneered the migrant road from the Missouri to the Columbia basin. Abandoning their canoes in canyons above Twin Falls, the explorers followed the Snake and Columbia to Astoria, Oregon, a walk of more than 800 miles. MacKenzie and Hunt were among the first white men to pass through the Shoshone trade centers in the Weiser area and near future Fort Hall.

Goods suchs as these — hatchets, pipes, gloves and other garments trimmed with trade beads — were exchanged for furs at frontier rendezvous.

Shoshone war club, about 1860s

of people living on the plains were the Snake River Shoshones, excellent equestrians who ranged far to the south. Among the Shoshone people were nomadic bands of Northern Paiutes, whom the white people called Bannocks.

Some Shoshone bands were known chiefly by their diets. There were salmon eaters who followed the migrating fish to Shoshone Falls, and buffalo eaters, sheep eaters and other salmon eaters who fished the wooded streams near Weiser. Another band of Weiser-area Shoshones, known as mountain sheep eaters, were skilled trappers and furriers who hunted in Hells Canyon and began an active trade with the Nez Perce to the north. They made high-quality buckskin. To the east, where the once mobile Shoshones now farm on a small reservation near Fort Hall, a large clan of buffalo eaters ventured into central Idaho and perhaps as far south into the Rockies as Jackson Hole, Wyoming. Numbering about 1,000 by the mid-1800s, they wintered and intermarried with members of the Bannock tribe. Sadly, their strategic location doomed them to conflict with the whites heading north from Utah. Although the Fort Hall Shoshones were efficient hunters, they were slow to acquire rifles and thus the tribe was seldom a match for heavily armed parties of whites.

The first Euro-Americans to encounter these tribes were the groups of adventurers from the United States and Canada who competed with Russians and others for control of the Pacific Northwest. One of the first explorers in the area was Andrew Henry, who crossed into eastern Idaho from Montana with some of Manuel Lisa's Missouri Fur Company trappers in the spring of 1810. The trappers made camp on what came to be called Henry's Fork of the Snake River; but unimpressed after a harsh winter at high elevation, they left the following spring.

Meanwhile, adventurer John Jacob Astor had sent Wilson Price Hunt and 65 trappers into the Rocky Mountains. Astor had sent others around the tip of South America, and the two expeditions were to meet at the mouth of the Columbia River. The overland Astorians spent nearly four months or more scrambling from the Tetons to the Columbia River. Hunt's

Hand-forged beaver trap, 1850s; opposite, Joe Meek, a trapper with Jim Bridger's Rocky Mountain Fur Company, spent a cold winter at the mouth of the Portneuf in January 1833.

main goal was to survive, but he had been ordered to investigate an overland route and locate potential post sites. The Snake River fascinated him. Near the Idaho-Wyoming border, Hunt named it Canoe River in the hope that the wide, quick-moving water would quickly carry the explorers to their destination. It was "light green … its banks covered with small cottonwoods." As Hunt advanced, "the river grew lovelier and wider." But Hunt and his men were deceived by the initial promise of the Canoe River. From Idaho Falls to American Falls to Caldron Linn and then from Twin Falls on to Shoshone Falls, they struggled with the forbidding environment. As the river sliced through the lava rock, it grew "constricted, full of rapids and bordered with precipitous rocks." There were delays, lost supplies and postponed fords. At least four trappers died.

Ironically, Washington Irving, the man who recounted this experience, never crossed the plain. Yet his descriptions in *Astoria* left the nation with vivid impressions. Writing from the Hunt party journals, he used phrases like "burnt and barren prairies," "barren rocky country," a "wide sunburnt landscape," a "dismal desert," "cheerless wastes and vast desert tracts." Irving's interpretation reinforced the popular perception of a wasteland along the Snake from the Rockies to the Cascades.

In 1813 a small group of Astor's fur-hunters, including Robert Stuart, returned to the East by way of the Snake, the Portneuf and the Bear rivers. Twenty-five years later this passage approximated the main overland route. Stuart's story, published by Irving and St. Louis newspapers, reported fish, beaver streams, varieties of game and rugged terrain. Stuart complained about "the detested shrubs" such as "sage, wormwood and saltwood" that covered "a parched soil, of salt, dust and gravel." Yet he marveled at the Snake River with its "terrific appearance," abundance of fish and thriving beaver population.

John James Audubon's American Beaver *from his wildlife study* The Viviparous Quadrupeds of North America *(1842–1854).*

These early expeditions opened an era of intense rivalry between British and American trappers. Two British enterprises, the North West Company and Hudson Bay Company, competed with Americans moving into the Snake River country. Donald MacKenzie, one of the original overland Astorians, went to work for the rival North West Company after Astor's fur scheme collapsed. Weighing nearly 300 pounds, MacKenzie threw himself into a venture that would lead him to travel throughout the plain. From 1818 to 1821, MacKenzie led British expeditions from the Columbia up the Snake and into the canyon lands of the high desert. His trappers, who found extremely fertile country, ultimately would beat a path across the plain that became part of the Oregon Trail. The Snake region, said MacKenzie, was "altogether … a delightful country."

International events accelerated the near ex-tinction of the Snake country beaver, whose high quality pelts were prized throughout the world for hats. The United States and Great Britain, unable to

Trapper rifle and powder horn, 1850.

settle on a boundary line in the Northwest, agreed in 1818 to joint occupation for 10 years. Americans called it the Oregon country. British trappers hoped to drive the Americans out by removing every beaver pelt. Perhaps, they reasoned, a beaver wasteland would force the Americans to shift their attention to California and the southern Rockies. Hudson Bay Company trappers Peter Skene Ogden, Alexander Ross, John Work and others traveled up the Snake River annually, living off the land, trapping as far south as the Great Salt Lake and California.

Ultimately the British extermination policy was only a partial success. The beaver population was decimated, creating a fur desert. But the devastation failed to deter American trappers. Jim Bridger, Osborne Russel, Warren Angus Ferris, Thomas Fitzpatrick and other adventurous Yankees crossed into the Snake River country from the upper Missouri basin during the height of the fur-trading era.

Highly romanticized figures, these American trappers epitomized the robust wander-lust of the young republic — but their legacy was grim. Some killed Indians freely. While the Indian warriors were not afraid to strike back at small bands of frontiersmen, the Shoshone learned to avoid head-on military confrontations with the well-organized Hudson Bay expeditions.

Jedediah S. Smith, a native of New York state, enjoyed better relations with the Indians than most. One of the great frontier explorers, he was a deeply religious man who always carried a Bible. In 1824 Smith joined some lost Iroquois trappers and guided them back to Alexander Ross' party of Hudson Bay trappers in central Idaho. Although Smith continued to wander, he, like other American trappers, preferred the mountainous forests.

By the late 1820s most of the beaver had been taken from the canyons and many Americans had retreated to the Wyoming highlands. The British remained on the plain, appearing to control the Snake River–Oregon country. The presence of trappers, not surprisingly, had a dramatic effect on the tribes. American and British trappers brought firearms, iron utensils, liquor and blankets — trade goods that transformed Indian life. Trappers also introduced terrible diseases. Measles, smallpox, cholera and other afflictions decimated entire western tribes, although the desert tribes were spared the worst of these epidemics. Near Bear River, however, a missionary reported a sickly clan of Bannocks ravaged by smallpox.

The trappers had another severe impact on the tribes of the eastern plain: the destruction of the immense buffalo herds. Although as late as 1833 American fur trader Benjamin L.E. Bonneville reported "immense herds" of buffalo near future Pocatello, the days were numbered for the great beasts. As the herds declined, so did the fortunes of the tribes that depended on them for survival.

Exploring the Snake. The "discovery" of the Snake River country was the process through which explorers, trappers and topographers redefined the land; from left to right: "King of the Northwest" Donald MacKenzie (1783–1851); mulatto trapper James P. Beckwourth (1798–1867?); and U.S. topographical engineer John C. Frémont (1813–1890).

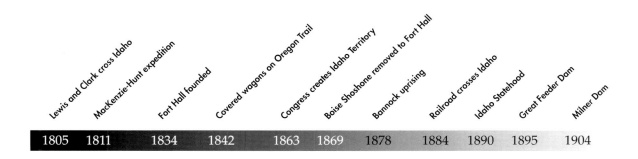

Lewis and Clark cross Idaho	MacKenzie-Hunt expedition	Fort Hall founded	Covered wagons on Oregon Trail	Congress creates Idaho Territory	Boise Shoshone removed to Fort Hall	Bannock uprising	Railroad crosses Idaho	Idaho Statehood	Great Feeder Dam	Milner Dam
1805	1811	1834	1842	1863	1869	1878	1884	1890	1895	1904

Map of the Territory of Oregon *(1844). The Democratic* Ohio Statesman *published this black map as a protest to the Henry Clay–John Quincy Adams compromise that honored British claims to the disputed Northwest.*

Alfred Jacob Miller's Setting Traps for Beaver *(1837).*

Beaver hat, about 1810.

The 1830s, a decade of revived American interest in the Oregon country, saw a shift in the fortunes of white traders and settlers on the Snake River Plain. One adventurous agent of change was Nathaniel Wyeth, a Boston ice merchant. In 1832 Wyeth took a caravan of trade goods to American trappers at Pierre's Hole rendezvous in Idaho's Teton Valley. By the time the shipment arrived, Wyeth had been cheated and beat out by a rival, the famous mountain man Thomas Fitzpatrick. The rendezvous exploded into a horrendous battle among trappers, Flathead, Blackfeet, Shoshone, Bannock and Nez Perce Indians. At least 21 men were killed.

Laden with supplies but without buyers, Wyeth proceeded on through south-central Idaho, down the Snake and eventually to the mouth of the Columbia. His main contribution to Idaho history was the construction of Fort Hall, overlooking the Snake River near the present-day site of Shoshone-Bannock tribal headquarters outside Pocatello. Completed in August 1834, Fort Hall was the first permanent American outpost in the wild region beyond the Continental Divide. Meanwhile the Hudson Bay Company, responding to American advances, built the Fort Boise trading post on the Snake at the mouth of the Boise River.

Wyeth helped Americans reassess the value of the Snake River country. A keen observer of geological features, he wrote about the basin's "strong volcanic appearance" and its "streams [that] occupy what appear to be but cra[c]ks of an overheated surface." Wyeth saw a future for agriculture. The basin, he said, was a potential oasis with rich soil. But excessive cold and lack of rainfall prevented the plants from "assuming a fertile character." Outmaneuvered by the British, Wyeth returned to Boston, and in 1837 he sold Fort Hall to his rival, the Hudson Bay Company.

Benjamin Bonneville, a captain on leave from the U.S. Army, also reached the Snake River country in 1832. Bonneville was the first to drive wagons with oxen into the basin, but the trip was painful and slow. Although the fur-trading venture he embarked on was a financial failure, Bonneville became famous through the publication of his *Adventures*, narrated by Washington Irving in 1836. Here the plain was "sandy and volcanic," "incapable

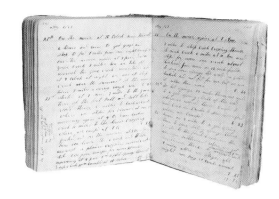

A. S. Davis trail diary (1853).

of cultivation," and having had numerous "volcanic convulsions." Near Fort Hall, the lower Portneuf was "rent and split with deep chasms and gullies, some of which were partly filled with lava." As far as Bonneville and Irving were concerned, the whole region "must ever remain an irreclaimable wilderness." Facing up to the harsh realities of climate and the decreasing price of beaver, Bonneville returned to his career in the regular army.

The golden era of the Snake River fur trade, which had lasted less than 50 years, came to an end. As fashion changed and silk hats replaced beaver, trappers left the mountains. Some served as guides for two new kinds of migrants: missionaries and land promoters. American trappers, the British could handle; evangelical ministers and speculators, they could not.

Missionary interest began in the 1830s with Jason and Daniel Lee, New England Methodists who traveled with Wyeth. Sent west by the Methodist Missionary Society, the Lees, however, preferred the well-watered green Willamette Valley of Oregon to the Snake River country.

The first women to cross the continent by wagon were members of another expedition. Narcissa Prentiss Whitman and Eliza Hart Spalding accompanied their husbands,

"Aug. 23rd On the moove again at 7 A.m. 6 miles to Steep Creek crossing thence to Sink creek 6 mies at 11 Am and stoped for noon one mule about tuckered, p.m. mooved on 12 miles further to some worm springs the water is quite brackish and not fit for use but we use it rather than move."

— Oregon Trail Diary

In 1887 Miss Amelia J. Frost came to Fort Hall to establish the first Presbyterian mission. Her work among the Shoshone continued until 1907 when her fragile health forced her to return east. The photograph at right captures Frost with her first group of schoolgirls.

Studio portrait of a Shoshone family from the Smithsonian archives, about 1869.

Missionary woman believed to be Narcissa Whitman.

Axle-grease graffiti, City of Rocks, a landmark on the overland trail.

Dr. Marcus Whitman and the preacher Henry H. Spalding, on a famous expedition outfitted in 1836 by the American Board of Foreign Missions.

It was a difficult and miserable journey. Reaching Fort Hall in the heat of summer, the early missionaries suffered under dreary conditions. Narcissa Whitman wrote: "Heat excessive. Truly I thought the Heavens over us were brass, and the earth iron under our feet." Elsewhere the party was "so swarmed with musquetoes [sic] as to be scarcely able to see." As they traveled across the plain, she noted "a species of wormwood called sage of a pale green, offensive to both sight and smell," but she also observed that "the country is barren and would be a sandy desert were it not for the sage." The pace of the journey was unforgiving. Twice during their journey across the plain, at Fort Hall and Fort Boise, they washed their clothes and relaxed.

These women, however, were bolstered by their missionary zeal. They had faith in the significance of what they had undertaken. Of her plight, Eliza Spalding wrote: "I have often spoken of the fatigue and hardship I have experienced on this journey, I have experienced many, many mercies which ought to dissolve my heart in thankfulness & cause me to forget the inconveniences I endured on the journey."

Like the Whitmans and the Spaldings, many of the 50,000 migrants who would cross southern Idaho in the next 25 years reached Idaho at the height of the summer heat. By August the lush grasslands had become a brown and blistering desert. "We got along tolerably well until after we passed Fort Hall," wrote one pioneer. To the west of the fort, the "grass was very carce [sic] nearly all the way down Snake River. Cattle began to give out and a great many died ... I was taken sick with the mountain fever ... our provisions gave out and we had like to have starved to death ... we had to kill our work cattle for beef poor as they were and eat them." Almost every journal confirmed that bleak assessment.

Another traveler remembered that when "the Snake River was reached and in fact before, the heat again became oppressive, the dust stifling and thirst at times almost

View from Camp Ground, August 22, 1849. *Halfway between Glenns Ferry and Bruneau on the "dry" south alternate trail.*

maddening. In some places we could see the water of the Snake, but could not reach it as the river ran in the inaccessible depths of the canyon." As the years progressed, grass became more scarce along the trail and many oxen and cattle perished. "One of our oxen died day before yesterday," a woman recorded, "and one of John's today. He has lost two, Fred one and Davie Love one." Two weeks later, she said that three more had died.

One bizarre account of the Oregon Trail was that of the religious mystic William Keil. Keil traveled with the body of his 19-year-old son preserved in alcohol in a lead-lined casket. Terrified by the lava, he claimed that the "seventh prince of all destruction" inhabited the desert. It was the devil's landscape: "hideous world, fearful roads, all grass poisoned, every day one to three head of cattle dying, a killing heat, nothing to see but the marks of death and destruction, the whole road marked by the graves and the bones of dead men."

Others, however, could appreciate the scenic splendor of the volcanic crescent. While the plain's eastern region was viewed as barren, the migrants became more enthusiastic as they moved northwest from the Bear River to the Portneuf, the Blackfoot and the Snake. That landscape was viewed more favorably. In 1849 James A. Pritchard noted that "the valley continues handsome and fertile" and "we came to a small creek and in one mile thereafter we found a splendid spring of water, that gushes from the base of the mountain. The grass continues fine, the mountain sides are covered with a stinted seeder [cedar]."

Once the pioneers left Fort Hall and began moving westward toward the verdant landscape of Washington and Oregon, the environment changed dramatically. As the trail approached Fort Hall, the migrants first saw the Snake. One traveler called it a "clear and beautiful stream of water. It courses over a pebbly bottom. Its width is about one hundred and fifty yards. It abounds in fish of different varieties, which are readily taken from the hook." Moving on, migrants saw the magnificent American Falls, then Salmon Falls, Thousand Springs and Three Island Crossing. Shoshone Falls and Twin Falls, two of the most breathtaking sites, were north of the great migration. Not until the era of the U.S. government surveys would the nation come to appreciate the natural wonders beyond the Oregon Trail.

To enjoy the desert leg of the journey, one traveler wrote, "a man must be able to endure heat like a Salamander, . . . dust like a toad, and labor like a jackass"; detail from Albert Bierstadt's Oregon Trail *(1869); left, toddler shoes, found along the trail.*

FEDERAL EXPLORATION

One of the first government explorers to rethink the prevailing attitudes about the intermountain region was John C. Frémont, a handsome captain in the U.S. Corps of Topographical Engineers. With orders to survey the route between the Platte and Sweetwater rivers in 1842, Frémont extended the expedition through the South Pass of the Rockies. The following year he set off for the Pacific with 40 men, among them a German cartographer, a free black from Missouri, two Delaware Indians and legendary guides Kit Carson and Thomas Fitzpatrick. The 1843 expedition made its way from Independence, Missouri, to Wyoming and Utah and along the Oregon Trail through Idaho from Fort Hall to Fort Boise and then on to Fort Vancouver and California.

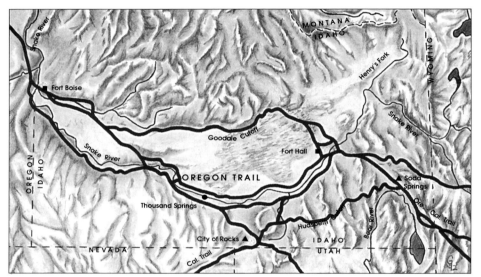

*Idaho's overland trails,
1841–1862.*

Back in St. Louis, Frémont published stirring accounts of his exploits (polished considerably by his wife and editor, Jessie Benton Frémont, the daughter of Missouri senator Thomas H. Benton). Captain Frémont's reports convinced many that the journey west was both feasible and practical. Stressing the beauty of the Snake River country, he overlooked the forbidding character of the lava desert. He noted "dark rocks" along "green and wooded watercourses, sunk in chasms." In spring, he imagined, "the contrasted effect must make them beautiful." Praising the area around Fort Hall and Fort Boise, he felt that the posts were near fertile and productive soil. However, like so many others, Frémont believed that the land between the forts was fairly worthless. "Beyond this place [Fort Hall] there does not occur, for a distance of nearly 300 miles to the westward, a fertile spot of ground sufficiently large to produce the necessary quantity of grain, or pasture enough to allow even a temporary repose to the emigrants." Yet the captain was quick to add a positive note: the tall grass was "nutritious" forage. Wheat might grow in the sage-covered soil. Thus Frémont, an outspoken expansionist, was one of the first to predict that human enterprise would transform the arid crescent. He would prove to be quite right.

Frémont's era was a time of territorial conquest and great ambition that Americans called "Manifest Destiny." In 1846, on the eve of the Mexican War, President James Polk's administration signed a treaty with Britain that ceded the Oregon country — the future states of Oregon, Washington and Idaho — to the United States. Oregon became a state in 1859. Idaho, torn away from Washington Territory and encompassing future Montana and most of Wyoming, achieved territorial status in 1863. Territorial reorganization brought a new, more vigorous wave of government exploration — boundary surveys, road and railroad surveys, fort-building expeditions, punitive campaigns against the Shoshone and quasi-scientific ventures. West Point–educated U.S. topographical engineers made a great show of science, but their knowledge of the Snake River country was secondhand and quite superficial. Many army reports were thinly disguised appeals for federal aid for a Pacific

railroad from the Great Lakes to Seattle. Focusing on northern routes, they stressed Idaho's beauty. Common soldiers sent to fight Indians in the Idaho desert were less optimistic. Of the 1,121 infantry and cavalrymen who fought the Shoshone at Bear River in 1863, about one-fifth, 238 men, deserted the frontier army before serving out their four-year enlistment.

With the reorganization of the topographical corps during the Civil War, the government shifted resources from military exploration to civilian expeditions. In 1868 civilian scientist Clarence King visited the Snake River with a party of the U. S. Geological Survey of the 40th Parallel. Drawn by reports of heavy coal (it was black lava), he camped at Shoshone Falls, watching it through the night.

> Intervals of light and blank darkness hurriedly followed each other. Tall cliffs, ramparts of lava, the rugged outlines of islands huddled together over the cataract's brink, faintly luminous foam breaking over black rapids, the swift, white leap of the river and a ghostly, formless mist through which the canyon walls and far reach of the lower river were veiled and unveiled again and again.

The Snake River Plain, once thought by migrants to be a barren and unforgiving land, was now becoming an exotic frontier, a basin of natural wonders. Ferdinand V. Hayden, a Smithsonian geologist and geographer with a flair for self-promotion, was another scientist who came to appreciate the plain. A native of Massachusetts who grew up in New York, Hayden spent 11 years surveying, studying and analyzing the West.

Scientific explorers like Hayden and King form an important bridge from the past to the future. They were the first to value the land apart from geopolitical considerations. It was no longer a territory to get past or merely an area to be acquired. "There seems to be no want of fertility in the soil of our western plains," Hayden insisted, "and when the two most

Frontier outposts, about 1849; left, Fort Hall; Fort Boise.

Clarence King's U.S. Geological Survey of the 40th Parallel above Shoshone Falls, 1868.

Mormon missionaries crossed the Snake River Plain in 1855; above, the Nauvoo Legion guarding the overland trail; right, detail from Lee Green Richards' 1945 mural study for the Idaho Falls temple, the first LDS temple in Idaho and the only one on the banks of the Snake; inset, the temple at night.

important conditions are favorable, climate and moisture, or water for the purpose of irrigation, then agriculture will be a success."

Thus Army exploration and the precision of science altered the popular perception of the Snake River country. Government explorers were quick to see the fantastic promise of irrigated agriculture. As they mapped and studied the Idaho desert, they were aware of another significant fact: white settlers were very close behind.

THE MISSION AND THE MINES

Despite the expansionist tone of the government surveys, most Oregon-bound pioneers still dismissed the Snake River country as dreary and inhospitable. In the 1850s and 1860s the impact of two events changed that popular perception: Mormon settlement and the rush for silver and gold.

Followers of the Church of Jesus Christ of Latter-day Saints, commonly called Latter-day Saints or Mormons, had trekked west from Illinois to Utah in 1847, fleeing religious persecution. The first Mormon expeditions to reach Idaho were part of a grander plan. Mormon leader Brigham Young hoped to pacify, baptize and Christianize native tribes from the Rockies to California. According to Mormon theology, the native Americans are a chosen people, one of the tribes of Israel, who deserve the blessings of Christianity. More important than the conversion of Native Americans, however, was the establishment of Mormon control over large tracts of land. In 1855 Thomas S. Smith led 27 missionaries on a 380-mile journey north through the Snake country, then part of Oregon Territory. On the Lemhi River, near the fork of the Salmon not far from where Meriwether Lewis had crossed into Idaho a half century before, Smith and his men built the Fort Lemhi mission.

Initially the mission prospered as the missionaries built a sawmill, gristmill and blacksmith shop and planted crops. Fort Lemhi ultimately took on the appearance of a permanent farming community, and relations between the mission and various groups of Indians were harmonious. More than 100 Indians were baptized, and later, young Mormon men were encouraged to marry Indian women to cement the friendship that had deepened.

Brigham Young inspected the site, and settlers flooded in from Utah — but a number of problems on the near horizon soon forced a Mormon retreat. Since 1849, when the California gold rush had diverted some of the western travel from the Oregon Trail, trouble had been brewing between Young, who was the territorial governor, and federal officials. Young used his religious and secular authority to circumvent the federal marshals; judges and officials did all they could to paint a gloomy picture of conditions in Utah. As Young scrimmaged with federal officials, President James Buchanan sent an army into the West to quell the so-called Mormon Rebellion.

Anti-Mormonism unsettled the Idaho mission. Other Americans in the area, mountaineers and traders, feared the territorial expansion of Mormonism. Indeed, Indian agents reported that the Mormons were offering ammunition to the Native Americans in order to fight the federal army. Rumors and threats circulated among Mormons and non-Mormons alike. At the center of their fears were the Native Americans. Of course, the Indians were told that the Mormons were after their land, that Mormons only converted them to make them peaceful. The mountaineers, who realized that the Army needed supplies and would pay cash to get them, persuaded the Indians to raid the mission.

In 1857 the clash came to a head when the missionaries entertained a band of Nez Perce. The visitors made off with 60 Shoshone horses. Local tribes, fearing an alliance

138

Missionary Thomas S. Smith, about 1855; below, Shoshone encampment, 1860s.

between the Nez Perce and the Mormons, protested outside the mission, demanding concessions and grain. In late February 1857 two missionaries were killed when Bannock and Shoshone warriors took most of the settlers' cattle. Outnumbered, the mission sent for help, and soon Brigham Young called the men back to Utah. The Mormons had learned a lesson. With water the land could be conquered and the soil would produce, but the Snake River Indians were not about to sit on their hands.

The Mormon "war" and the breakup of the Lemhi mission only delayed colonization. In 1860 a party of 16 colonists founded a Mormon community just north of the Utah border at Franklin, Idaho. They built a fort, a store, log cabins, two mills and ditches for irrigation. As Mormon settlement spread through Shoshone country, cattle crowded out elk and bison and atrocities were exchanged between the settlers and the Indians. Soon several bands of Shoshone were in open revolt. When the Franklin settlers called for protection in January 1863, Col. Patrick E. Connor and about 200 of his California Volunteers responded with terrible force. At dawn on January 29, the U.S. troops raided a Shoshone camp, methodically shooting into the crowd. About 400 Shoshone men, women and children were slaughtered in the bloodiest massacre on record in the history of the Northwest frontier.

To the west, a gold rush was putting new pressure on area tribes. With the discovery of gold on August 2, 1862, at Grimes Creek near Idaho City, thousands of miners poured into

the mountains, and homesteaders along the Boise River cleared land to supply the mines. Some of these first pioneers were pro-Dixie Confederates from Missouri — refugees of the Civil War. On July 4, 1863, Maj. Pickney Lugenbeel established a fort on a hill overlooking the crossroads between Idaho City and the rutted Oregon Trail. Three days later a few farmers met with Lugenbeel to lay out a town, the future city of Boise.

Soon the Boise Valley was not only a marketing center; it was an agricultural power-house — Idaho's largest producer of wheat, oats, hay and cattle. Boise City became the new territorial capital, with a prosperous Main Street, waterwheels and ditch irrigation. By the 1870s the capital was the plain's largest city, with about 1,000 residents.

The natives of the Boise Valley did not share the wealth. Dispossessed by farming and ranching, their winter encampments disrupted by placer mining in the Snake River canyon, the Boise-area Indians avoided contact with whites. When U.S. Indian agents worked out a series of treaties with Chief Pocatello and other Shoshone leaders in 1863, the Indians living in the Boise and Bruneau areas were excluded, their rights transferred to other tribes. At last on April 16, 1866, territorial governor Caleb Lyon signed a treaty that promised a Bruneau Canyon reservation. The treaty never made it through Congress. While the natives living in the Boise area all but abandoned the valley, small bands of North-ern Paiutes raided farms in the Owyhee-Malheur area near the Idaho-Oregon line. In response, whites demanded punitive expeditions that made little distinction between peaceful and militant tribes. Campaigning mostly in the winters of 1866 and 1867, federal troops out of Fort Boise combed the Snake River canyon, driving some Bannocks and others deep into the Oregon desert. Some Boise and Bruneau clans found refuge in the foothills; others were relocated by the government. In 1869 about 1,150 Shoshone and 150 Bannocks abandoned their ancient homeland for reservation life at Fort Hall.

Although isolated skirmishes continued, the pitched battles ended after a small band of rebellious Bannocks returned to the Fort Hall reservation in 1878. Settlers were now free to homestead, harness the rivers and tip the balance of nature that once sustained Indian lands.

Detail from Arm Hincelin's Main Street, Boise *(1864).*

IRRIGATION AND SETTLEMENT

Idaho irrigation began with Henry H. Spalding, who in 1838 dug a ditch from the Clearwater River to his dying garden. Thirty years later, while Boise settlers were experi-menting with ditches and waterwheels in dry Ada County, Mormon pioneers spread irrigation to the Lemhi, Cache, Bear Lake and Malad valleys. Each region found its own way to finance these projects. In the Boise Valley, for example, irrigators relied on private investors, many from the East. On the eastern plain the Mormons joined church-sponsored community irrigation ventures. The Great-Feeder Dam near Rexburg, a Mormon project completed in 1895, fed one of the world's largest and most successful networks of gravity canals. Aridity forced Idaho irrigators to radically revise the Anglo-American tradition of water allocation. Heretofore, the English common law doctrine of riparian rights was the standard in 19th-century America, even in the arid West. Riparian rights held that landholders whose lands were adjacent to a stream were entitled to as much water as they desired. In arid climates this meant that only lands next to waterways could be successfully developed. But miners and Mormon irrigators immediately saw the need to transport water far from the river, and thus the Snake River pioneers endorsed the water-rights doctrine of prior appropriation. The earliest water users had the strongest legal claim.

In 1890 the state constitution, claiming the state's water belonged to the citizens of Idaho, said that applications for water had to be made through the state. A series of court cases strengthened the doctrine of prior appropriation. State law also set priorities for Idaho water usage, with domestic needs first, followed by agriculture and manufacturing. A farmer or a manufacturer who wasted water was guilty of a misdemeanor. Since so much of early irrigation in southeastern Idaho began prior to the formulation of Idaho law, numerous legal tangles ensued. Wisely, the state encouraged the creation of irrigation districts, a western innovation that enabled landowners to organize, assess taxes and float bonds for dams and canals.

In 1899 a group of midwestern Swedes organized an early irrigation district at the New Sweden colony west of Idaho Falls. Caldwell-area irrigators created a district in 1901, and by 1916 there were 30 such organizations. While the districts had the power to mediate some local conflicts, a history of passionate fights in and out of the courtroom still sprung from disputes over water. In the eyes of many farmers, especially during dry years, water thievery was a grave crime.

As Idahoans competed for water, the farmers of the Boise Valley looked as far as New York and London for the money to build canals. In 1882 a group of New York and East Coast investors organized the Idaho Mining and Irrigation Company, a Boise canal company. Soon mining engineer Arthur D. Foote planned an enormous project that Boiseans called the New York Canal. Seventeen feet deep and 27 feet wide, it would run 75 miles and feed more than 5,000 miles of lateral ditches. Foote projected that nearly a half million acres of arid land could be brought under cultivation.

Like many engineers unfamiliar with steep canyons and desert conditions, Foote far underestimated the challenge. Construction stalled and two Philadelphia investors tried to rescue the venture with a smaller link to the Nampa area, the Phyllis Canal. When the Idaho Central Railway offered to take over construction, the new owners refused to sell. Eventually the New York investors reestab-

Diversion Dam on the Boise River, completed in 1909.

"DREAMERS WE ARE"

Arthur Foote

Mary Hallock Foote

In 1883, while still abed after the birth of her second child, Mary Hallock Foote received a letter from her engineer husband Arthur De Wint Foote. He was in Idaho inspecting new silver mines for prospective investors, and Mary was anticipating news of his return to their family in New York's Hudson River Valley. Instead, she read of his astonishing plans to design and build a huge irrigation canal and diversion dam in southwestern Idaho. That idea, now known as the New York Canal, left Mary cold.

After years of traveling through the West and Mexico with her husband on his mining projects, Mary was no stranger to hardships, but the canal project meant banishment to "darkest Idaho." Idaho, she said, was "thousands of acres of desert empty of history."

But Mary relented. Arriving in Boise in 1884, Mary, a novelist, kept the family afloat as an essayist and illustrator for *The Century Illustrated Monthly Magazine*. An instant celebrity, she entered the polite, backwater society of "well-meaning" yet "dowdy and dull provincial ladies [who] kept making courtesy calls."

Still Mary had a novelist's eye for the land — its hypnotic blankness, its powerful beauty. During lean and frustrating years in the Boise River canyon, a time when her marriage struggled and Arthur's water scheme seemed to be failing, Mary was still able to love the land for what it was. "I forsee [*sic*] the time," the writer confessed to a friend, "when I shall long for [the West] and be homesick for the waste of moonlight, the silence, the night wind and the river! There is nothing that will ever quite take its place! Dreamers we are, dreamers we always will be, and what is folly and vain imaginings to some people is the stuff our daily lives are made of. And there are thousands like us! If there never had been there would be no great West."

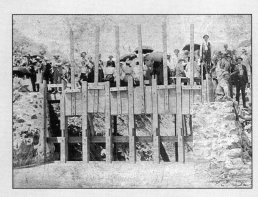

Opening ceremony, New York Canal, 1900.

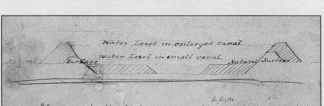

Arthur Foote's sketch for the proposed Pocatello Canal, 1893.

Union Pacific pamphlet, about 1911; right, Twin Falls alfalfa farmer, about 1910.

lished control, and with the help of Denver builder W.C. Bradbury, water reached Nampa in 1891. Three years later nearly $400,000 — over half of it Bradbury's own money — had been poured into the main canal. Still, 14 critical miles remained unfinished and bankruptcy loomed. In 1894 Bradbury hired future senator William E. Borah, one of Idaho's great legal minds. Borah arranged for an Ada County sheriff's sale and managed to obtain both the New York and Phyllis canals for a rock-bottom price, about $184,000.

By the time the New York Canal opened in 1900, Foote, Bradbury and the rest of the early developers were out of the picture. Two competing groups finished the canal's construction. Ultimately Charles Fifer organized the New York Canal Company, which appropriated water to shareholders in proportion to the number of shares they owned. Assisted by federal funds authorized under the 1902 Reclamation Act, a canal system was finally completed. The finished product was a strange canal. Broad in places and narrow in others, it brought water to 38,000 acres by 1906. Eventually the U.S. Reclamation Service enlarged the canal and built a diversion dam.

The lesson of the New York Canal was that community enterprise and private effort were seldom enough. Large-scale reclamation depended on outside investors, engineering vision and, increasingly, state and federal aid.

Agriculture on the Snake River Plain also depended on an efficient railroad network. Railroads first reached southern Idaho when the Mormons of Cache Valley brought the Utah Northern to Franklin in 1874. The Utah Northern, which reached Garrison, Montana, by 1882, cut through the heart of the Fort Hall reservation. Also in 1882, Union Pacific's Oregon Short Line came to Idaho. When company officials decided to connect their transcontinental line to the Northwest, they proposed a line from Granger, Wyoming, across Idaho to Oregon where it would connect with the Oregon Railway and Navigation

NAMES ON THE LAND

Bannock Country, Massacre Rocks, Spud Butte, Railroad Canyon — the place names of the Snake River country are clues to its vibrant past.

The frontier era survives in names of Indian leaders (Pocatello, Targhee Creek), trappers (Payette, Portneuf) and federal explorers

Walters Ferry

(Fremont County, Hayden Peak). Walters Ferry and Three Island Crossing recall the Oregon Trail.

The boom years of early statehood are well represented in the names of railroad officials (Drummond, Burley) and irrigation tycoons (Kimberly, Buhl).

Caldwell takes its name from a Kansas senator who speculated in western lands.

Place names are also ethnic references (Danish Flat, Geneva), racial slurs (Nigger Creek, Jap Creek), battle sites (Battle Creek, Soldier Mountains), flora and fauna (Camas County, Salmon Falls), and include a thousand interesting stories. Malad River was named by Donald MacKenzie after his trappers feasted on beaver and became ill (*malade* meaning "sick" in French). Owyhee, a 19th-century spelling for Hawaii, referred to South Pacific islanders who disappeared while trapping the southwestern Snake River country during the winter of 1819.

Buhl

ALFALFA.
TWIN FALLS, IDAHO.

BISBEE PHOTO.

Company. Initially the suggested route ran parallel to the Oregon Trail. However, railroads did not need frequent access to the river, so the rail bed was built north of the canyon.

The railroad raised money through its aggressive subsidiary, the Idaho and Oregon Land Improvement Company. Company directors such as Pittsburgh financier Andrew Mellon and Kansas senator Alexander Caldwell profited from advance notice of station sites. The land company surveyed town sites, but without a guaranteed water supply farmers were reluctant to settle.

Publicist Robert Limbert with his papier-mâché model of Shoshone Falls, part of his award-winning display at San Francisco's Panama-Pacific International Exhibition, 1915; above, flag from a 1911 state immigration and labor report; above right, Miss Ida-ho, Harvest Queen, sells Idaho's abundant resources, 1911; opposite page, postcards pictured a bountiful land adorned with natural wonders.

Turning from land sales to mining, the Oregon Short Line made quick profits when a branch line was built from Shoshone to the Wood River gold rush at Hailey. Tracks were laid quickly, and by February 1884 the line stretched into Oregon. It spanned the mighty Snake four times and originally bypassed the territorial capital in Boise because of the downhill grade into the city. Including the spur to Hailey, the Union Pacific constructed more than 500 miles of track on or adjacent to the Snake River Plain. The Union Pacific's Short Line was a communication revolution for Utah, Oregon and southern Idaho. In 1884

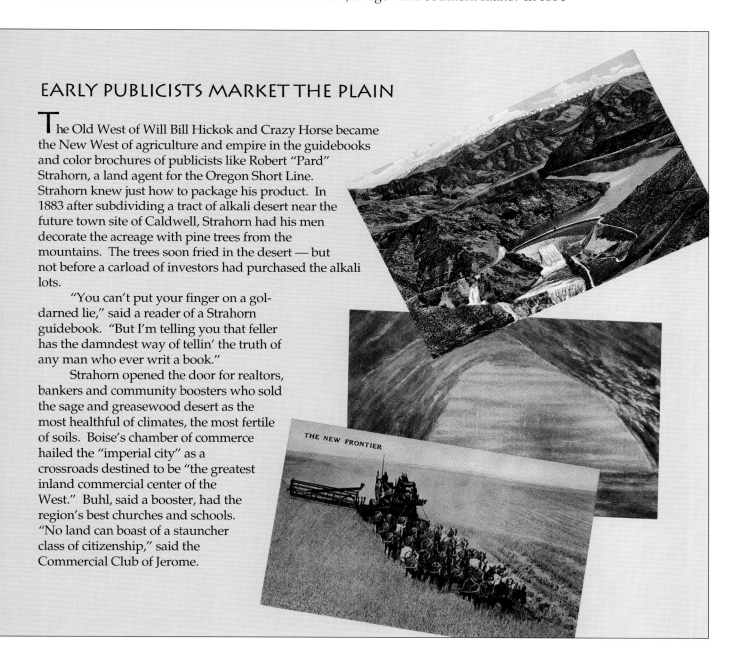

EARLY PUBLICISTS MARKET THE PLAIN

The Old West of Will Bill Hickok and Crazy Horse became the New West of agriculture and empire in the guidebooks and color brochures of publicists like Robert "Pard" Strahorn, a land agent for the Oregon Short Line. Strahorn knew just how to package his product. In 1883 after subdividing a tract of alkali desert near the future town site of Caldwell, Strahorn had his men decorate the acreage with pine trees from the mountains. The trees soon fried in the desert — but not before a carload of investors had purchased the alkali lots.

"You can't put your finger on a gol-darned lie," said a reader of a Strahorn guidebook. "But I'm telling you that feller has the damndest way of tellin' the truth of any man who ever writ a book."

Strahorn opened the door for realtors, bankers and community boosters who sold the sage and greasewood desert as the most healthful of climates, the most fertile of soils. Boise's chamber of commerce hailed the "imperial city" as a crossroads destined to be "the greatest inland commercial center of the West." Buhl, said a booster, had the region's best churches and schools. "No land can boast of a stauncher class of citizenship," said the Commercial Club of Jerome.

THE NEW FRONTIER

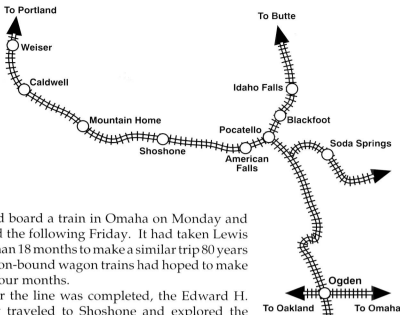

Far left, Boise-bound passenger train; left, ticket to Caldwell, 1908; above, Union Pacific route map, about 1925.

a passenger could board a train in Omaha on Monday and arrive in Portland the following Friday. It had taken Lewis and Clark more than 18 months to make a similar trip 80 years before. The Oregon-bound wagon trains had hoped to make the same trip in four months.

Shortly after the line was completed, the Edward H. Harriman family traveled to Shoshone and explored the surrounding area, looking for ways to increase interest in Idaho. Here was the seed of what would become Sun Valley, the world-famous ski resort founded by Averell Harriman in the 1930s.

Numerous other small railroad feeder lines were constructed throughout Idaho. When the Utah and Northern temporarily merged with the Oregon Short Line, Pocatello became the "Gate City" to the West Coast and Montana. Now a railroad community, Pocatello would become a center for non-Mormons in an area heavily populated by Latter-day Saints. Bumper crops in 1919 brought Mexican field hands to the Pocatello area, and by 1920 366 blacks, many of them railroad workers, lived in the city.

Railroads allowed agriculture to spread and opened the Snake River Plain to more profitable homesteading. Under federal homestead laws, farmers or ranchers could, for a nominal fee, gain title to public land if they were willing to live on their tract for five years and make improvements. But most of the laws limited the homestead to 160 acres (320 acres for a husband and wife). Sen. William Borah summed up the homesteaders' challenge: "The government bets 160 acres against the entry fee of $14 that the settler can't live on the land for five years without starving to death." Even when the Desert Land Act of 1877 expanded the homestead to 640 acres, the law required successful irrigation within three years. That expense was well beyond the common farmer. Thus, federal homesteading did not bring large populations to desert climates, but public land programs did pave the way for some surprising developments.

One creative approach was the 1894 Carey Act, a program based on the notion that reclamation and private enterprise could profit with some help from the states. Proposed by Wyoming senator Joseph Carey, the act provided up to 1 million acres of federal land to any western state that would be willing to supervise the acres. The land could be sold in parcels as small as 40 acres, but at least that many acres had

Union Pacific Railroad promotional literature sold Idaho as a desert oasis. From 1907 to 1911, William Bittle Wells, chief publicist for the U.P.'s Bureau of Community Publicity, flooded the Northwest with promotional literature — pamphlets, postal folders, magazine articles and color brochures. Small-town promotional pamphlets were brightly decorated with ornate borders and classical motifs.

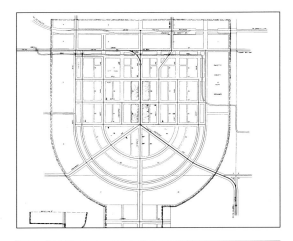

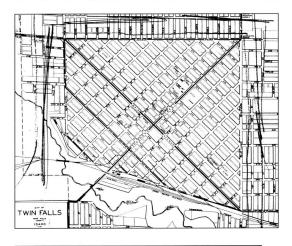

Grids and futuristic innovations helped town builders tame the desert. Arc-shaped New Plymouth, top, a utopian colony, drew money and inspiration from the 1894 National Irrigation Congress. Oakley, middle, typified the wide streets and rectangular symmetry of Cassia County's early Mormon settlements. Twin Falls, bottom, was platted in 1903 by the reclamation corporation that invested in Milner Dam.

to be brought under irrigation. The settler could obtain a patent on the land once it was irrigated. Later amendments allowed the state to place a lien on the land to protect private capital. Investors now had 10 years to complete an irrigation project. Construction companies sold water rights to the individual farmer, and the state sold the land for 50 cents an acre — half down, the rest upon final proof of improvement. When the irrigation system was completed, the irrigators would operate the system. The state worked with entrepreneurs to plan and supervise the disposal of Carey Act lands. Most states were unable to take advantage of the Carey Act, but Idaho did, and the Twin Falls project became the national showcase of successful desert reclamation.

The Twin Falls project began to transform arid south-central Idaho into the productive farming regions of Cassia, Gooding, Jerome, Lincoln, Minidoka and Twin Falls counties — the Magic Valley. Frank Riblett, a pioneer surveyor, saw the promise of this valley; so did engineer John Hayes and farmer John Hansen, but these visionaries lacked capital. Even though some farmers filed appropriation notices and surveyed their canal lines, the lean 1890s were poor years to ask for money.

It remained for land developer Ira B. Perrine, a gifted promoter from Indiana, to take up the work of the early surveyors. In 1900 Perrine persuaded Stanley Milner of Salt Lake City to invest $30,000 in an irrigation survey. After some initial financial setbacks, Perrine and Milner won the support of Witcher Jones, a mining broker and financier. Perrine also earned the trust of a Pennsylvania steel millionaire named Frank H. Buhl. Impressed by the Magic Valley after a tour with Perrine, Buhl returned to the East; aided by associate Peter Kimberly, Buhl's new Twin Falls Land and Water Company began issuing bonds.

In 1903 the Twin Falls company promised to develop more than 244,000 acres under Carey Act provisions. If the state sold the land and the company sold the water, both would profit. The terrain, after all, was not especially rugged. An impressive diversion dam named for Milner was completed in 1904. It fed a canal 10 feet deep and 80 feet wide. Perrine's dogged persistence proved to be vital: it took the sight of water flowing through the canal to sell the land. Buhl, Kimberly and their general manager, Walter Filer, had almost pulled out their support after the initial land sale attracted only about a dozen people. But by the end of 1905, Twin Falls was an agriculture boom town.

Buhl was not interested when Perrine asked him to finance a project north of the Snake, so Perrine convinced the Kuhn family of Pittsburgh to join him in a northside development. However, significant problems were to come. The land north of the river was too porous for a storage reservoir, and although the large tract was ultimately sold with access to Milner Dam, the promoters sold more land than they could supply with water. Kuhn money also financed a troubled Carey Act project on Salmon Falls Creek southwest of Twin Falls. Again, the project was oversold and the developers were never able to deliver sufficient water.

In 1913 a bad situation turned worse when the collapse of irrigation schemes in Colorado and elsewhere sent banks into a panic. Although the Kuhn projects were solvent, the financial crisis touched off a recession, and the vast Kuhn empire — land, railroads, coal mines and banks — was hurt nationwide. The northside project was completed only after Perrine and canal manager Russell E. Shepherd persuaded the U.S. Bureau of Reclamation to increase the supply of water to the Magic Valley by building a dam across an outlet of Jackson Lake.

Although hundreds of thousands of Idaho acres were brought under irrigated cultivation through provisions of the Carey Act, not all projects enjoyed the enormous success of the Twin Falls venture. The state had to involve itself in troubled Carey Act projects near King Hill below Twin Falls, the one on Salmon Falls Creek and the Big Lost River project near Arco. Idaho Gov. Moses Alexander, a German-born immigrant, was the moving force behind irrigation reform during the First World War. Alexander, the first elected Jewish governor in the United States, cut the size of some irrigation districts and expanded others with water from federal projects. He also proposed that the state pay court costs for any settler who sued a construction company that failed to deliver water. These and other innovations help Idahoans revive Carey Act projects at a time when the West was pressuring Congress to build dams to flood the desert.

"Loaded and starting to hoist" — Arrowrock Dam construction, 1912.

GOVERNMENT AND WATER

Federal reclamation changed the politics of the Snake River country as the water altered the land. President Theodore Roosevelt, with his stewardship theory of government, was attracted to proposals for large-scale federal projects. In 1902 Roosevelt joined Nevada senator Francis Newlands in pushing for the National Reclamation Act, or Newlands Act, a law that helped to fund dams and canals through the sale of western lands. The act created a regional bureau of the Interior Department, the Reclamation Service. Renamed the Bureau of Reclamation in 1923, the agency had a tremendous impact on southern Idaho. Reclamation engineers pioneered high-dam construction at Minidoka, American Falls and Palisades. From 1915 to 1934, the bureau's dam at Arrowrock on the Boise River was the tallest in the world. By the 1960s a string of federal projects from Jackson Lake to the Owyhee River made the upper Snake and its tributaries one of the West's most developed river basins. The federal government also used Idaho as a testing ground for rural electrification programs. Today, hydropower remains one of Idaho's chief exports.

One of the earliest and most ambitious federal reclamation projects was the Minidoka Dam. Located on the Snake between Burley and American Falls, about 35 miles upstream from Milner Dam, the Minidoka project began in 1904 and was completed in 1906. A gravity canal irrigated land north of the dam while the south side relied on a series of pumping stations. By 1913 when the power plant opened, the Reclamation Service had invested more than $6 million. The successful Minidoka project spurred the rise of irrigation districts and politically savvy water-user associations that lobbied for additional projects. Congress in 1914 won grassroots support by allowing local districts to collect user

WRANGLERS AND TURKEY TENDERS
VERSATILE WOMEN OF THE SNAKE RIVER PLAIN

From clearing fields to selling crops, a woman's work on a pioneer farm was whatever had to be done. Women cooked, made textiles and managed finances, and because the Idaho homestead often included several tillable plots, the farm wife was frequently responsible for her own herds and crops. Women, said historian Richard B. Roeder, were "the economic linchpins" of the family homestead, the key to a venture's success.

During the early years of Ada County, from 1869 to 1890, at least 25 women filed for independent homesteads while many others claimed land jointly with their spouses. At age 20 Elizabeth Onwiler was one of the original pioneers of the Meridian area. Permeal J. French of Boise and Hailey, a future university dean, homesteaded and farmed in the 1890s to supplement her meager income as a rural schoolteacher.

Women raised cash on the farm by selling butter, eggs, wool and other products. They also herded turkeys. In the Caldwell area, where women were organizing poultry co-operatives as early as 1910, homesteaders used turkeys to launch the hugely successful Idaho-Oregon Turkey Growers Association, a marketing organization. The association, made up mostly of women, shipped carloads of dressed turkeys as far as Florida and Maine. By 1926 southern Idaho turkey ranching was a $1.5 million business.

Pioneer women also found ways to ranch cattle and sheep — even if it meant dressing like a man. Little Jo Monaghan, a Snake River wrangler, kept her sex a secret to the grave. Short and stocky with a high squeaky voice, Little Jo came west

Rose Stienmeyer Falk, an Idaho debutante, about 1875.

Payette apple packers, about 1907.

Quilt panels, Weiser, about 1915.

with the Idaho gold rush and soon earned regional fame as an excellent horseman. In Owyhee County, where homesteaders herded Texas cattle as early as 1869, ranchers like Jean Heazle began wearing trousers for purely practical reasons — skirts were awkward, even dangerous.

While most ranch women retained their traditional roles as cooks and caretakers, some were the brains of large operations. In the 1920s Anna Joyce managed a big herd at Sinker Creek near Murphy. Idaho's "Horse Queen," Kitty Wilkins of Glenns Ferry, was a fearless rider who broke wild mustang's and ran one of the state's largest horse ranches.

Cowboy Jo, 1904.

Turkey tender, about 1910.

IDAHO'S "SHEEP KING"

One of the first to run sheep on the north side of the Snake River in the Minidoka desert was Scottish immigrant James Laidlaw, who arrived in Idaho in 1891. He built the first roads in the area.

Laidlaw brought the first bands of sheep into the Carey area in 1895. The dry climate and range conditions — grass, wildflowers and weeds — proved very favorable for raising sheep. Laidlaw quickly became one of the state's most prominent sheep producers and at one time ran nearly 30,000 sheep in central Idaho. His home range was the hills north of Carey extending into the present Muldoon region.

Known as the "sheep king" of Blaine County, he began working with Rambouillet and Lincoln crosses to develop the rugged Panama breed, a larger, good meat- and wool-producing, but durable and range-hardy sheep adapted to Idaho's high altitude and mountainous terrain. One of the most useful attributes of the breed, especially in the rugged, open range lands of southern Idaho, is its tendency to flock together, discouraging predators. Each winter, the Panamas made an 85-mile trek from the Laidlaw Ranch at Mul-doon to their lambing sheds northeast of Rupert. The Panamas, one of only two recognized sheep breeds developed in the United States by private breeders, received its name after being introduced to sheep producers for the first time at the 1915 Panama-Pacific Exposition in San Francisco.

The Laidlaw family, for some years with partner Robert Brockie, another Scot, continued to raise the Panamas, selling them throughout the United States and South America. In 1951 a Panama registry was formed. Laidlaw was actively engaged in the sheep industry until his death in 1950. After his widow died in 1958, their sons owned and managed the ranch. A portion of Laidlaw's ranch at Muldoon is still owned by his descendants.

fees on behalf of the Interior Department, and in 1921 the Reclamation Service sponsored legislation to ease restrictions on federal loans.

The apparent success of federal reclamation brought a land rush to southern Idaho that ignited disputes among farmers and launched hundreds of ambitious schemes. On the south side of the Minidoka project, for example, there was often a severe shortage of water, which fueled a demand for more storage upstream. Reclamation engineers responded with a timber crib dam in western Wyoming that turned several natural pools into Jackson Lake. The impact of drought, which seemed to occur about every five years in southern Idaho, created increased pressure on Congress. As the demand for Idaho crops expanded during the boom years of World War I, the problem of how to administer the Snake River became

Right, light bulb, about 1910; far right, early Idaho Power Company logo.

the mind-boggling task of deciding who would get what — the process known as water adjudication. Vicious, occasionally violent, disputes arose when downstream irrigators accused upstream farmers of holding back water.

The severe drought of 1919 brought these disputes to a head. Crops valued at $15 million were lost, and the Snake dried to a trickle at Blackfoot. In this atmosphere of crisis the Idaho Reclamation Association was born. Organized in Pocatello in 1919, the reclamation association brought together many local groups, and in 1923, the composite organization elected a powerful governing board, the so-called Committee of Nine. Then in 1924, upper-basin farmers went on to fuse local groups into the world's largest irrigation district, District 36, a working example of irrigation democracy.

Meanwhile the Committee of Nine fostered an atmosphere of enlightened cooperation. It convinced farmers that the normal flow of the Snake River had been overextended by the continual opening of new projects. The committee then hired a Colorado engineer, F.T. Meeker, to study water storage. Meeker said Minidoka and Jackson Lake reservoirs were inadequate for the present needs, let alone the future demands. More reservoirs would guarantee water year-round. The accuracy of Meeker's report, the success of the Committee of Nine and a spirit of cooperation were demonstrated during the drought of 1924 when farmers with prior rights in the upper basin donated water to irrigators near Twin Falls, allowing crops to mature in the Magic Valley.

One of the next likely dam sites was American Falls, a location with several major obstacles. More than just a construction matter, a dam at the site required that the entire town of American Falls be relocated, land had to be purchased from the Shoshone-Bannock Fort Hall Reservation, three miles of Union Pacific track had to be rebuilt and a million-dollar arrangement had to be made with Idaho Power. Idaho Power, a local company which played a significant role in rural electrification, ran a generator at the base of the falls. Albert B. Fall, the new secretary of the interior appointed by President Warren G. Harding, visited American Falls in 1921. After learning that Minidoka irrigators were behind in their government payments, Fall backed away from federal funding for the American Falls dam. He could not see the sense in expanding the federal role even if the Reclamation Service did want to build a dam.

Major dams and reservoirs on the Snake River Plain.

While the American Falls project was in limbo, the Idaho congressional delegation, led by Sen. William Borah, lobbied the Interior Department. Borah promised that the state would work with the irrigators and Idaho Power and outlined the concept of the single irrigation district. In 1923 Fall promised funds,

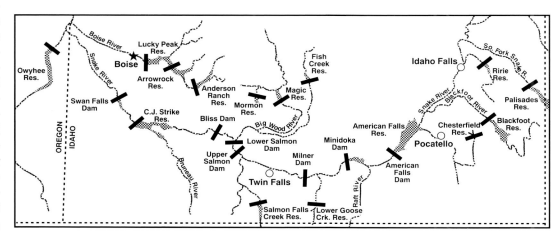

Laura Gipson's Cornfield *(1984).*

Hay knives, 1880s–1930s.

but the farmers would have to raise part of the money. They did and the $8 million, 25-mile reservoir was filled in 1927. The project impounded 2 million acre-feet of water and irrigated about 1 million acres, guaranteeing, for a time, a steady supply of water to the Magic Valley.

Although the American Falls project brought greater security to farmers downstream, irrigators in the upper valley still lobbied for more storage capacity. The row crops of the eastern plain — beans, sugar beets and potatoes — needed water late in the summer. But no irrigation system could regulate the erratic snowfall that fed Idaho's water supply. Boom years gave way to falling prices, drought and the Great Depression: Snake River farmers lost more than $7 million in 1935 alone.

Idahoans in Congress continued to lobby for water. By 1959 the Bureau of Reclamation completed a huge dam and reservoir at Palisades just west of the Wyoming border. A multipurpose facility, it stored water for reclamation, hydropower, recreation and flood control. Soon power companies were locking horns with irrigators and conservationists over a series of dams on the lower Snake in Hells Canyon. The battle for Hells Canyon forced Idaho Power to subordinate its water rights to farmers in the upper basin. The agreement set the stage for years of litigation and a new era of conflict between irrigators, industrialists, federal water agencies and environmental groups.

In the 1960s these same interests battled again over the Teton Dam. The Teton River flows into the Snake north of Idaho Falls. Opponents claimed that the site was geologically unstable and therefore unfit for a dam, but Congress authorized the project in 1962. After a decade of loud debate, the Bureau of Reclamation began construction in the early 1970s. On June 6, 1976, the nearly completed dam collapsed. A great wall of water swept through the valley, killing 11 people and flooding the communities of Wilford, Sugar City, Salem, Hibbard and Rexburg. The torrent destroyed headgates, canals, farm buildings, hundreds of businesses and thousands of homes — causing at least $1 billion in damage. In all probability, the Teton Dam will not be rebuilt, but some Idaho farmers still hope to revive the project.

THE ETHNIC LANDSCAPE

From the success of Minidoka to the disaster of Teton Dam, the checkered history of big water projects transformed agriculture and the face of the land. Reclamation also radically altered the state's culture, bringing growth and ethnic diversification. Chinese-Americans, once Idaho's most populous Asian minority, were among those influenced by changes in agriculture. The Chinese-Americans made the slow transition from mining and domestic employment to farming on land rejected by others. In early 20th-century Boise, for example, a group of Chinese farmers lived on Government Island and planted large vegetable gardens along a strip of flood-prone land.

When federal restrictions cut off Chinese immigation, many Japanese families from Hawaii and Japan filled the demand for field hands and backbreaking railroad work. By 1907, Japanese crews in Idaho Falls and Blackfoot were harvesting sugar beets for about 50 cents a ton. Idaho's Japanese, unlike the Japanese-Americans closer to the Pacific, were not forced into federal relocation centers during World War II, but there was wartime hysteria and blatant discrimination. Some anti-Japanese laws remained into the 1950s.

Many ethnic European groups moved into the Snake River country, adding their own rich traditions to the cultural mix. Greeks immigrated to Pocatello. Malad had a Welsh settlement. Slavs joined Mexicans and blacks in the Union Pacific's Snake River rail yards, and a group of Czechs homesteaded near Buhl. German immigrants were some of early Boise's most prominent citizens. A Jewish community was established in Boise as early as 1869, and in 1882 there were so many Italians in Shoshone that people called it "Naples."

The Snake River Plain also would become home to a large population of Euskaldunak people, or Basques, from northern Spain and southern France. First-generation Basques

Mexican-American migrants work a Sunnyslope apple harvest. More than 100,000 migrant and seasonal workers, many of them Hispanics from Texas, are indispensable to the state's agricultural economy.

Basque-American weight lifter with granite ball, 1980s. Southwestern Idaho has one of the largest concentrations of Basques outside France and Spain.

Amalgamated Sugar Co. advertisement, Camp Minidoka yearbook, 1943.

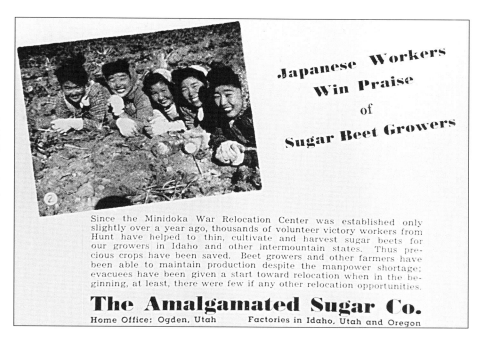

found employment on Idaho's rangelands — the men working chiefly as sheep herders, the women often staying in town. Many members of the second generation left ranching to become construction workers, shopkeepers, hotel owners and stockbrokers. Boise had a Basque neighborhood known for its boarding houses and hotels. By 1922, when federal quotas reduced immigration, there were perhaps 3,000 Basques in the Snake River country.

Today one of the plain's fastest-growing ethnic groups is Mexican-American. Since the era of the gold rush there were Mexicans in Idaho working as cowboys, miners and field hands. The 1920 census listed 1,125 Idaho residents of Mexican birth. During World War II the state began tapping federal funds to import migrant labor to southern Idaho under the bracero program. Mexican braceros supplemented a large agricultural work force of schoolchildren, housewives, Indians, about 1,000 Jamaicans, German prisoners of war and more than 2,000 Japanese-American "evacuees" from Camp Minidoka, a bleak relocation center. Inferior pay, inadequate housing and contract violations forced the Mexican government to cancel the bracero program in 1945. A year later striking Mexican farmworkers shut down four migrant camps near Nampa. Although conditions remained deplorable, food processing revived postwar agriculture, and the demand for seasonal workers continued to rise. The 1960 census indicated that there were 3,341 Mexican-Americans in Idaho, but some migrant workers were not included in the official count. The 1990 census listed more than 20,000 Hispanics in and around the farming centers of Nampa, Caldwell and Weiser. In 1991 the Idaho Migrant Council estimated more than 58,000 Hispanics living year-round in the Snake River country.

Thus the remarkable Snake River Plain is a culture of contrast and change. Waves of immigration have transformed southern Idaho as deeply as water and agriculture have remade arid terrain. Now an agribusiness center, a hydropower factory and a tourist destination, the plain is no longer the raw frontier; yet enduring disputes over nature's resources suggest a residue of frontier determination, and our society's need to subdue the Idaho desert recalls a history of conquest and big water projects — a page from our turbulent past.

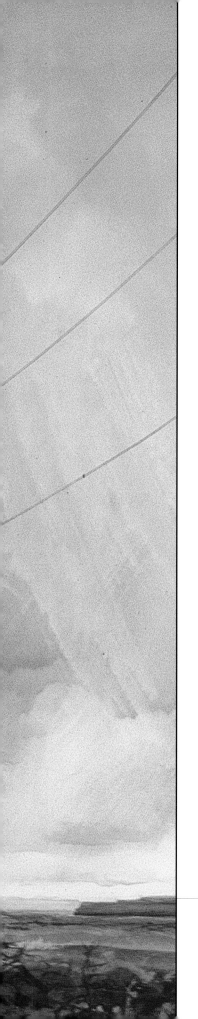

What we call Man's power over nature turns out to be a
power exercised by some men over others
with nature as its instrument.

— *C. S. Lewis*

The Abolition of Man, 1943

POLITICAL
LANDSCAPE

IN THE EARLY DAYS of irrigated farming prior to the advent of
a legal, orderly system of dividing up the mighty Snake for watering
crops, pioneering farmers had to cope with the hazards of physically
defending their water rights. These were lawless, white-knuckle times —
a dog-eat-dog atmosphere in which irrigators had to fend for themselves.

BY STEPHEN STUEBNER
WITH JOHN FREEMUTH

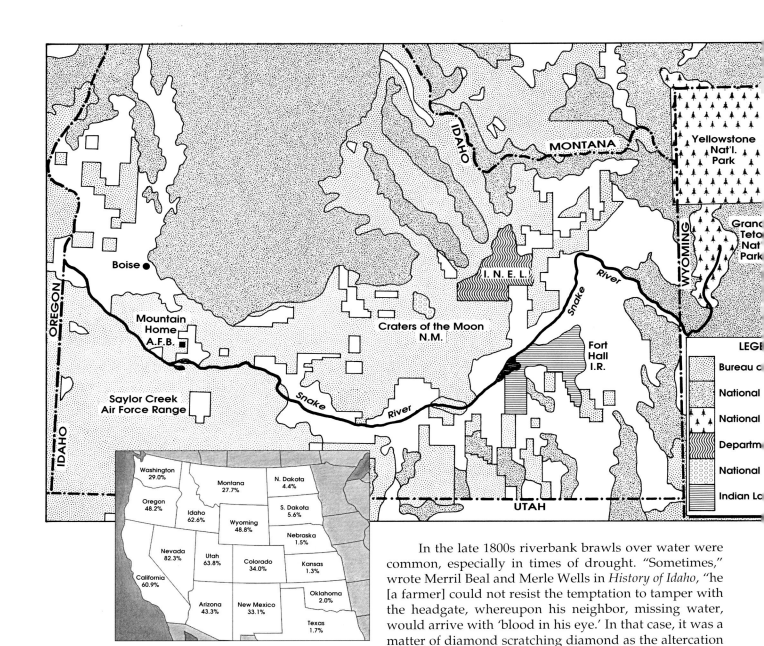

Washington 29.0%
Montana 27.7%
N. Dakota 4.4%
Oregon 48.2%
Idaho 62.6%
Wyoming 48.8%
S. Dakota 5.6%
Nebraska 1.5%
Nevada 82.3%
Utah 63.8%
Colorado 34.0%
Kansas 1.3%
California 60.9%
Arizona 43.3%
New Mexico 33.1%
Oklahoma 2.0%
Texas 1.7%

Federal land ownership for 17 western states; previous page, artist John Killmaster's portrayal of the Snake River country; detail, Idaho Statehouse cupola.

In the late 1800s riverbank brawls over water were common, especially in times of drought. "Sometimes," wrote Merril Beal and Merle Wells in *History of Idaho*, "he [a farmer] could not resist the temptation to tamper with the headgate, whereupon his neighbor, missing water, would arrive with 'blood in his eye.' In that case, it was a matter of diamond scratching diamond as the altercation mounted."

Even in the 1930s, some 30 years after Idaho had adopted the prior-appropriation doctrine — first in time, first in line for doling out water for irrigation and municipal purposes — heated water disputes occurred. In the Big Lost River Valley, for instance, farmers with long-held water rights on the fringe of the Big Lost got fed up with Utah transplants who sought to divert their water to the uncultivated sagebrush plain via a new canal. The transplants wouldn't take "no" for an answer, so Big Lost farmers eventually fetched some dynamite and blew up the canal, putting an end to that scheme. As recently as 1973, a Glenns Ferry farmer shot an intruder caught in the act of stealing his water.

Indeed, to a farmer on the Snake River Plain, water rights are more valuable than money. "Water is to Idaho what coal and oil are to the great states of Pennsylvania, Utah and Wyoming. Water is to Idaho what copper is to Montana. Water is to Idaho what natural gas is to Oklahoma and Texas," noted Alex O. Coleman, St. Anthony farmer and director of the Idaho Reclamation Association in 1955.

Wind River I.R.

Public land ownership. About three-fifths of southern Idaho is rangeland administered by the U.S. Bureau of Land Management. The lakes and reservoirs, covering about 275,000 acres, occupy more than twice as much land as the cities and towns.

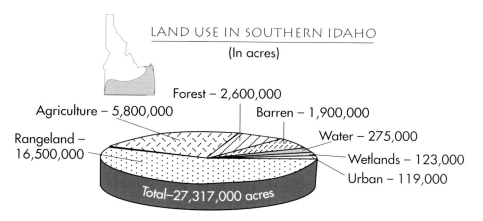

LAND USE IN SOUTHERN IDAHO
(In acres)

Forest – 2,600,000

Agriculture – 5,800,000

Barren – 1,900,000

Rangeland – 16,500,000

Water – 275,000

Wetlands – 123,000

Urban – 119,000

Total–27,317,000 acres

Figures for Ada, Bannock, Bear Lake, Bingham, Blaine, Bonneville, Butte, Camas, Canyon, Caribou, Cassia, Elmore, Franklin, Fremont, Gem, Gooding, Jefferson, Jerome, Lincoln, Madison Minidoka, Oneida, Owyhee, Payette, Power, Teton, and Twin Falls counties.

For most of the 20th century, farmers toiled to develop the plain and to put every drop of water to use — be it for growing crops, producing hydropower or raising rainbow trout in hatcheries. Nearly all of the irrigation canals were installed on a local basis around a stream or reservoir. No one realized in the early stages how the various irrigation networks might fit together, but eventually, they coalesced into a fascinating whole. Former Republican governor Robert Smylie articulated the state's intent for the Snake to members of the U.S. Senate in the mid-1950s when Congress was contemplating building a high dam in Hells Canyon. "We are seeking — we always have been and we always will — the ways and means of developing, comprehensively, every drop of water that tumbles from the snow packs of the Snake River watershed," Smylie said. "And when we have used that water, whether to help grow a potato or to turn a turbine, or both, then, and *only then*, will we willingly send it flowing into the canyons below Weiser to help develop still another empire further west."

In essence, the idea was to apply the bountiful waters of the Snake to the land, store surplus flows in dozens of reservoirs, and send the rest into the Snake River Aquifer, which also receives substantial runoff from sources such as the Big Lost River near Arco. Put another way, it was a use-it-or-lose-it doctrine designed to keep Idaho's water in Idaho for Idahoans to use.

Over the last 130 years, thousands of energetic and enterprising farmers across the Snake River Plain have brought about 3.6 million acres of dry sagebrush into bloom. Farmers tap into bountiful water supplies provided by deep snowpack in the high reaches of the Teton Mountains on the Idaho-Wyoming border and dozens of other mountain ranges encircling the plain to raise the nation's largest potato crop, the second-largest sugar beet harvest, and the third-largest volume of barley, hops and mint, among others. Agriculture, including ranching, is Idaho's largest private industry, yielding $3.1 billion in cash receipts

Idaho sugar beets, a $176 million industry, is the state's fourth-largest crop.

in 1991 and accounting for 27 percent of the state's gross product. Combined with food processing, agriculture makes up more than one-third of the state's gross product.

Today, however, a number of forces have come to bear on the Snake River that threaten to change the balance of the irrigation network that feeds the agricultural economy. In the 1990s, like never before, farmers will be competing with other interests, and with each other, for water — this time in court, in the Idaho Legislature and U.S. Congress, maybe even on the riverbank. The 1990s are a watershed period, one that will have a long-lasting impact on how political and environmental forces will likely reshape the flow of water on the Snake River Plain.

These forces include:

❑   An increasing number of interests are seeking a greater share of the Snake River for recreation, to maintain wildlife habitat or, in some cases, to protect public health. As it is now, few remedies exist for those seeking to capture a greater share of the Snake River, short of buying surplus water from irrigators. Some experts say court decisions based on the public trust, which takes all uses of the water into consideration, could force farmers in the future to free up more of the Snake for recreation and other uses. To some, this may seem like an idle threat. But the public trust can be a lethal legal weapon, according to University of Michigan law professor Joseph L. Sax.

Sax sees the public trust doctrine as being the supreme law on water. "Let me start by saying that as a matter of legal analysis, the holders of existing water rights are in deep trouble," the professor told a water symposium audience in Berkeley, California, in 1988. "The law is that water flows to benefit those uses that advance the contemporary public interest. No private right may stand in the way of that flux and reflux of water rights. Since the public interest, as now perceived, demands the retention and augmentation of in-stream water supplies, that is the way the water is going to flow."

❑ The federal listing of four species of Snake River salmon as threatened and endangered under the Endangered Species Act may give federal agencies a legal hook for dipping into previously untouchable irrigation water in the upper Snake. It is widely held that eight federal dams on the lower Snake and Columbia rivers in Washington and Oregon are responsible for killing more than 90 percent of the salmon on an annual basis. But federal

Sorting Idaho apples; below, sprinkler irrigation in Canyon County.

agencies charged with operating the dams have refused to make significant changes to address fish mortality; responding to downstream interests in Washington and Oregon, the agencies have preferred to call for increased output from the upper Snake to increase salmon survival. This issue will be a constant tug-of-war between Idaho, or "upstream" interests, and the more politically powerful Oregon and Washington "downstream" industrial interests.

❑ Five species of snails that reside in the middle Snake River downstream from Twin Falls also have been listed as threatened and endangered, setting up another possible conflict for upper Snake water. In large part, federal authorities indicated a need to protect the snails due to heavy pollution in the middle Snake region between Milner Dam and Thousand Springs. Joe Nagel, director of the Idaho Division of Environmental Quality, has called this river segment "an open sewer." While nearly all of the river is diverted at Milner into irrigation canals, a myriad of industrial pollutants are discharged into the Snake, now just a trickle, below the dam, creating massive algae blooms and depleting oxygen supplies for fish, aquatic insects and snails. Pollution sources include more than 500 feedlots and dairies, fish waste from more than 120 fish farms, hundreds of agricultural return flows, and treated waste from municipal water and sewer treatment plants. Under the Clean Water Act, public interest groups

IDAHO'S FAMOUS POTATOES

The genius of a hardworking teen-aged pig farmer named J. R. Simplot planted the seeds that produced a multimillion-dollar potato industry in Idaho, earning the state the moniker that it displays on its license plates — "famous potatoes."

Simplot, now a billionaire and the undisputed potato baron of the world, built his empire not just by growing potatoes on the farm, but by finding ingenious ways to add value to the spud: he produced freeze-dried potatoes for U.S. soldiers in World War II; he turned potato waste products into cattle feed, a spinoff benefit for his extensive cattle feedlot business; and perhaps most brilliant of all, he hired scientists in the late 1940s to create frozen french fries.

Today, the Simplot Co. is the world's second-largest producer of frozen potato products. Other sizeable companies have jumped into the action as well, such as Ore-Ida Foods, Universal Frozen Foods, Sun-Glo of Idaho, Magic Valley Foods, Inc. and Lamb-Weston, all located in southern Idaho. Together, these firms serve fast-food chains such as McDonald's, Burger King and Arby's and many frozen-food manufacturers.

Simplot dropped out of school in Declo at the age of 14 to run a pig farm. By the time he was 17, he had sold his hogs for a tidy profit of $7,500, purchased a team of horses and rented 160 acres of farm ground near Burley. He planted potatoes.

Today, southern Idaho farmers raise the nation's largest crop of potatoes, or about 30 percent of the nation's yield. Many of the spuds are sold to potato processing companies across the Snake River Plain. Between the farm gate and french fry, experts estimate that processors add nearly six times the value of the raw spud, from about 5 cents per pound to 30 cents per pound. In 1992 the Idaho potato was a $526 million business, the biggest cash crop in the state.

Potato harvest near Driggs; right, 1949 potato license plate.

could attempt to crack the zero-flow requirement at Milner Dam to help cleanse the middle Snake.

❏ In addition to these forces, the state is in the midst of adjudicating thousands of water rights for the entire Snake River Basin, a process in which the state must weigh claims of a variety of interests — irrigation, public health, wildlife and recreation. It potentially pits farmers with older, senior water rights against those with junior rights and in-stream irrigators against more recent groundwater pump irrigators. Nearly all experts agree that the Snake is overappropriated, meaning there are more farmers with water rights than there is water available, which at times leaves streambeds dry in the late summer months. Once water supplies become scarce, watermasters allocate the water on the basis of seniority — farmers with the oldest rights get served first. Because of the complexity and value of irrigation on the Snake River Plain, the adjudication has been jokingly referred to as the "attorneys employment act"; it is expected to roll on for decades.

Beyond the validation of individual farmers' water rights, the adjudication also will settle claims from the state and federal government for "reserved" water rights on public lands. This means new claims for water rights connected to federal lands, such as Deer Flat National Wildlife Refuge near Nampa, could conflict with farmers' historic water rights. If, for example, the U.S. Fish and Wildlife Service can defend the need for nearly doubling the Snake River's flow at Deer Flat to protect nesting geese and ducks on 86 islands, it would force the state to shut off water supplies to most of the high-lift irrigation tracts between King Hill and Caldwell. That is because Deer Flat would have a priority date of 1937 (the date of the refuge's establishment) for its reserved water right, and the high-lift tracts have priority dates in the 1960s. All of these factors and more make farmers in the Snake River Plain mighty nervous about the future.

Jumping the dunes — dirt bikes and all-terrain vehicles complicate the balancing act between preservation and recreation on Idaho's public lands.

As we have seen in a review of millions of years of life on the Snake River Plain, time never stands still — the wild creatures, the landscape and the people are in a constant state of flux. In the 21st century, things will be no different. Instead of farmers battling with other farmers for water, they will be forced to recognize that they have to find a way to share the Snake with other public interests, or the competing parties will seek relief through the courts, the Idaho Legislature, the Congress or other forums.

To Bliss native Peter Bowler, a specialist on the evolutionary biology of the middle Snake, the federal listing of the snails and the salmon indicate something is inherently wrong with the current management of the Snake River. These species are the veritable canary in the coal mine, Bowler says, indicating a danger to all forms of life in the food chain, including humans.

"We have all kinds of nationally significant resources in the Middle Snake, and now we have a chance to preserve and enhance them," Bowler said. "All of this is a really magnificent union of natural heritage and the public interest."

    Conflicts over water on the Snake River Plain reflect a change in public attitudes in the 1990s toward natural resources in general. When 19th-century pioneers toiled to tame the rugged landscape to make a living in farming, ranching, logging or mining, the concept of "preserving" resources for public recreation and scenic splendor was rarely considered. Where once the largely rural population of the Snake River country focused on bread-and-butter issues like economic growth and jobs related to resource use, today, Idaho's population is comprised of more and more people who harbor a strong environmental ethic and favor policies that could result in hardship for farmers, ranchers, loggers and miners.

    Surely the answer will lie somewhere in between. Over time, a new balance could be struck that gives stronger deference to the environment and the fragile natural resources of the plain, while fostering a greater understanding of the needs of Idaho's traditional industries. For if we review the progress of the past decade, we find several examples of how spirited conflict has inspired cooperative solutions, examples that could set a more conciliatory foundation for the future.

Fishing the Henry's Fork; above, green drake fly; below, power plant under construction at Island Park Dam.

THE HENRY'S FORK

Under the distant shadow of the Teton Mountains, the Henry's Fork of the Snake River glides through grassy meadows at the Harriman Ranch, near the tiny town of Island Park. Deep in the recesses of this cobalt blue stream, hefty rainbow trout hunker down in dark pools, casually awaiting the next meal. It is late June, and aquatic insects like the green drake mayfly are hatching on a warm summer morning. The newly hatched drakes float on the placid stream, becoming easy meals for the awaiting trout. Anglers fishing the river eye the unmistakable metallic glow of the drake's deep-green body. They tie a replica to the end of their fly lines.

On a day like this, anglers from as far away as New Zealand flock to the Henry's Fork to take advantage of the phenomenal green drake hatch. If an artful angler presents the green drake just perfectly, a 25-inch rainbow will zoom up from the stream bottom and — boom! — take the monster fly with a vengeance. "Set the hook!" the guide shouts, and the reel drag whines as the robust fish flees upstream. The fight is on.

Considering all the grandeur of the
Henry's Fork, and a strong national con-
stituency for protecting the stream's world-
class values, many were surprised in 1992
when the Idaho Legislature balked at im-
posing a ban on new dams, irrigation diversions and streambed mining on 146 miles of
the basin.

But the legislative fracas over the
Henry's Fork turned out to be the environmental donnybrook of the 1992 session. The
bill set up a classic clash in conservative
eastern Idaho between third- and fourth-
generation farmers and ranchers and anglers and environmentalists over the future
of the Henry's Fork Basin. Citizen debate
focused on what should be the state's future
policy for the area — whether or not to
continue an unmanaged blend of irrigation
use and hydroelectric development along
with trophy trout fishing, rich wildlife diversity and scenery. Would the state allow
new hydro projects to move forward that
could jeopardize the Henry's Fork values?
Or would the state shift the balance to favor
the stream's natural values at the expense
of development?

A power struggle ensued between the
old and the new, between continued development and resource protection, between
agriculture and tourism.

To farmers whose ancestors settled the
region, tourism is no substitute for hard-
core development, such as a new hydroelectric dam. "You can't build a tax base on
tourism," cautioned James Siddoway, a
Fremont County farmer and former chairman of the Fremont County Commission.
To boost the economy, county commissioners supported building a gravity-fed
hydroelectric project at Mesa Falls, a 320-
foot cataract. Bob Lee, former chairman of
the state Water Board, wanted to build a
4.5-megawatt hydro project at the
Yellowstone irrigation diversion on Fall
River. Lee promised that once the project
was paid off in 15 years, he would donate
40 percent of the revenues, or $400,000

annually, to Fremont County for public schools. The idea of additional development on the Henry's Fork, particularly Mesa Falls, was tough for others to accept. "I'm so mad, disgusted, so infuriated at what's going on," fumed Steve Christianson, a Jefferson County dentist, during a legislative public hearing. "The Henry's Fork doesn't just belong to the commissioners of Fremont County, and it certainly doesn't belong to developers."

Proponents of tourism argued that agriculture is important, but other assets need to be developed to diversify the region's economy. In Fremont County, unemployment had run nearly 10 percent in the early 1990s, about twice the state average. The poverty rate in the county also ran about double the state average. On the other hand, Yellowstone appears to be gaining a foothold in Fremont County, which butts up against the western edge of Yellowstone National Park. A 1992 economic study of the recreation use in Harriman State Park found that about 3,000 anglers infused $1.57 million into the Island Park economy during a 52-day period. "This proves that there is a major economic impact from those fishing this one small section of the Henry's Fork. Imagine the contribution fishermen must make throughout the basin!" observed Jan Brown, executive director of the Henry's Fork Foundation.

As it turned out, the Idaho Senate easily passed the Henry's Fork protection plan 32-10; but the Idaho House of Representatives killed it by a 44-40 vote. The House seized on two details in the plan: (1) three miles of protection for the Teton River could have prevented the rebuilding of Teton Dam, which failed in 1976, killing 11 people and causing an estimated $500 million in damage; (2) the plan encouraged retiring farmers to reserve their irrigation water for in-stream purposes. The second point was the straw that broke the camel's back.

Environmentalists had persuaded the state Water Resources Board to include this water rights language to benefit rainbow trout and endangered trumpeter swans, both of which had been killed by low river flows during the winter months. But farmers saw the conversion language as a red flag. It is one thing to protect the status quo, but it is quite another to encourage farmers to relinquish their coveted water rights.

Farmers work about 321,000 acres of irrigated cropland on the banks of the Henry's Fork between St. Anthony and Idaho Falls to grow mainly grain, potatoes and hay. Irrigation consumption here is abnormally high — with some farmers using four to six times the normal amount of water to raise a crop — as they pour large amounts of water down unlined canals and croplands to bring the groundwater level up to the root zone of the crops. This is called subirrigation. Due to high water consumption, farmers and ranchers in the basin consume nearly half of the average outflow of the Henry's Fork each year, or some 700,000 acre-feet out of 1.4 million acre-feet. Much of the water used here for growing crops seeps underground, eventually increasing the yield of the Snake River Aquifer at Thousand Springs and benefitting farmers downstream.

"The consumptive use of water is the very thing that makes our state thrive," Rep. Tom Loertscher, a Republican farmer from the Idaho Falls area, said during the debate. "If this bill has no protection for preserving our consumptive water rights, I'll vote no."

Ashton historian Dean Green noted after the vote, "History shows that opposition to anything that agricultural water users support is interpreted as a vote against 'mother-hood.'" But Green, who argued that early farmers in the basin better understood the dual benefits of the Henry's Fork, blasted the agrarian lawmakers for protecting themselves at the expense of the river. "Their need is only evident in their desire to serve themselves," he wrote in a guest opinion in the Idaho Falls *Post-Register*. "In the process, they draw the

WATER CONSUMPTION
(per capita gallons per day)

Idaho	22,200
Wyoming	12,200
Montana	10,500
Nebraska	6,250
Colorado	4,190

Idaho, sparsely populated and dependent on irrigation, uses more water per capita than any other state; below, canal headgate.

Cause and effect: collapsed Teton Dam and the flood cleanup in Sugar City.

lifeblood out of that beautiful river on the pretext of growth and agricultural need. They close their eyes to the recreational value the Henry's Fork gives to our souls. The Henry's Fork offers immeasurable value experienced by many seeking an emotional and recreational uplift from the everyday stress of life."

Perhaps fearing that killing the Henry's Fork bill would be a political liability, incumbent lawmakers quickly drew up an alternative measure that would place a temporary moratorium on development in the Henry's Fork Basin until a compromise could be reached. Senate President Pro Tem Mike Crapo, R-Idaho Falls, who was running for — and eventually was elected to — Congress in the 2nd District, authored the bill. It passed easily.

"I've fished the Henry's Fork all of my life, and I'd never do anything to hurt that river," Lynn Loosli, an Ashton farmer and rancher and Republican representative, told the House, urging members to support the interim protection bill. Sen. Laird Noh, R-Kimberly, branded the legislation "interim re-election protection for those who didn't vote for the Henry's Fork plan in the first place."

Meanwhile, back in the Henry's Fork Basin, several events occurred during the summer of 1992 that would boost the chances of protecting the Henry's Fork in the next Legislature. On July 11, a hydroelectric project under construction in the Fall River, a Henry's Fork key tributary, had an accidental breach. Water from the unlined canal seeped beneath a large pipe and broke through the canal into the Fall River, sending tons of sediment downstream. Gov. Cecil Andrus ordered an immediate halt to the project.

Environmentalists quickly used the incident to point out that if the Legislature had banned hydro projects like the one on Fall River, the beloved Henry's Fork would not suffer such a heavy blow. Then it got worse. In late September, the U.S. Bureau of Reclamation drained Island Park Reservoir near the head of the Henry's Fork at the request of the Idaho Department of Fish and Game. Fish and Game wanted to kill trash fish in the reservoir with Rotenone, a quick-acting toxin typically used for such operations. But the reservoir quickly emptied, and the Henry's Fork carved a channel through the muddy impoundment for two weeks, dumping more tons of silt into the river. Fish and Game was delayed in conducting its fish-kill operation. Once the agency applied the poison, it shut off the Henry's Fork at Island Park Dam, preventing the chemical from oozing into the main stream below. But it took longer than usual to detoxify the reservoir, leaving it solely up to the Buffalo River downstream to recharge the dry Henry's Fork channel.

Clearly, none of these calamities would have been prevented by bank-to-bank protection of the Henry's Fork by the Legislature. But to Jan Brown, executive director of the Henry's Fork Foundation, the events signaled the immediate need to protect the Henry's Fork from any more disasters. "It's a watershed in crisis," Brown warned.

THE NORTHSIDE CANAL CO.

Aerial view of Milner Dam showing diversions into the Northside and Southside canals; right, an Idaho watermaster.

The mighty Snake River winds across the eastern plain to a point near Burley, where it flows into Milner Dam, the dividing line between the upper and middle Snake.

Here the river splits in two and pours into one of two large canals, the Twin Falls Canal to the south and the Northside Canal. These two canals serve thousands of farms in the Magic Valley, bringing the former desert soils into bloom with a plentiful water supply. Under the Carey Act, Congress lured western pioneers with cheap land rates to take a chance on the rough and rocky turf and turn it into productive farmland. Farmers who took the dare in the Magic Valley—an area commonly referred to today as the "breadbasket" of Idaho agriculture — cashed in on a big bonanza.

Today, the bountiful waters of the Snake pour down the Northside Canal all summer long, serving about 2,000 farms and 160,000 acres on the north side of the Snake River. Here farmers raise dozens of crops such as potatoes, beans, sugar beets, corn, peas, grass seed, alfalfa hay, wheat, oats, barley and onions. Ted Diehl, manager of the Northside Canal Co. for a quarter century, said the canals extend for 1,100 miles just on this side of the Snake all the way to King Hill, where the waters dump into Clover Creek and, eventually, back into the main river.

The Northside Canal Co. is a nonprofit, non-taxable entity supervised by a nine-member board of directors. It was formed in 1907, well before many dams were built on the Snake River.

In the complex world of irrigation, canal companies and irrigation districts oversee the appropriation of water on the Snake River Plain. Farmers purchase water on the basis of shares. For the Northside Canal, three watermasters and 21 ditch riders make sure that each farmer gets his or her proper share.

Conversely, some irrigated lands are controlled by an irrigation district, also a non-profit organization, which generally provides water from federal reservoirs and divides it equally among its members. Each district holds title to water rights for its farmers. A canal company, though, may have some water rights, but other rights would be controlled by farmers themselves. Other kinds of canal companies provide water for farmers, but the farmers hold title to their own water rights.

Irrigation districts and canal companies are quite powerful under Idaho law: they wield the power of eminent domain, and they have the ability to assess fees to cover the costs of doing business.

In today's world, Diehl says, the life of a canal company manager is complicated by endangered species, hydroelectric projects, hydrogeologic studies, droughts and other factors. "It always seems that we're worried about some threat to our water system. But in Idaho, water rights are kind of etched in stone. I think we'll win most of the battles in court."

Small towns such as St. Anthony, right, are caught in the crossfire over the development of the Henry's Fork; middle, Gov. John V. Evans, flanked by James Bruce of Idaho Power (sitting left) and Idaho Attorney General Jim Jones, signs the 1984 Swan Falls Agreement that challenged the doctrine of prior appropriation by setting minimum flow requirements for the Snake River at Swan Falls; below, Henry's Lake, a state park, is also a privately maintained storage reservoir for Fremont County irrigators.

A survey of about 3,000 anglers that summer indicated the fishing in the Harriman reach of the Henry's Fork was "fair" to "poor." More than half the anglers attributed the fishing decline to poor water management and inadequate wintertime flows, both of which were blamed on agriculture.

The Water Resources Board revised the Henry's Fork protection plan to remove the two key points of earlier opposition: they revised the protection for the Teton River to avoid precluding a rebuild of the Teton Dam, and they removed any references that encouraged the conversion of irrigation water rights to instream flows. The board also added more potential hydroelectric sites that could, at some time in the future, be developed by irrigation districts or independent developers. But Lee and Fremont County dropped any active interests in their hydro projects. By the time the plan was ready for the 1993 Legislature, it had the support of the eastern Idaho farmers. Based on public input, the Water Board added nearly 50 miles of streams for protection to bring the total to 195 miles.

Henry's Fork advocates thought the board's plan was too weak. Along with 13 other environmental groups, they crafted a Henry's Fork protection plan for 279 miles. Key differences between the two competing plans were that Henry's Fork advocates wanted to protect several tributaries of Henry's Lake, the lower Fall River and the lower 50 miles of the Henry's Fork itself. Clearly, backers of the Henry's Fork 279 plan knew that their bill did not have a chance in the 1993 Legislature. House members saw the bill as being so radical that Henry's Fork advocates could not find anyone from eastern Idaho to sponsor it. When the Water Board's plan came up for consideration in the House Resources and

Conservation Committee, Democrats tried to amend the bill to conform with the 279 plan. That motion died for lack of support. Sherl Chapman of the Idaho Water Users said the Henry's Fork advocates were playing with fire by trying to amend the Water Board's bill. "If they had sent this bill to the amending order, we would have taken a butcher knife to the bill and gutted it," he said.

The Water Board's plan passed the House 70-0. Henry's Fork 279 advocates tried to amend the bill again in the Senate Resources and Environment Committee, but fell short by two votes. The Water Board's plan passed that chamber by another unanimous vote, 35-0. Andrus signed the bill several days later, calling it the "high water mark" of the 1993 session. "They don't make places like the Henry's Fork anymore, and it is our responsibility to be wise enough stewards to protect its waters for future generations," the governor said during the bill-signing ceremony. "This international treasure now has as its protectors environmentalists, farmers, sportsmen, irrigators and the rest of the people of Idaho."

Over a period of two years, Idahoans had come together on a controversial environmental issue and worked together on a resolution agreeable to most parties. Building on that greater understanding, the same parties that opposed each other in the 1992 Legislature over the Henry's Fork decided to form a new Henry's Fork Watershed Council in the fall of 1993. The Fremont-Madison Irrigation District and the Henry's Fork Foundation are co-chairing the council in hopes of developing more holistic management strategies for the Henry's Fork Basin, and to bring more than 30 government agencies and other players to the same table.

At the time of its inception, the council was one of the first in the western United States to bridge the traditional gap between farmers and environmentalists. Already, officials from both groups say they have a greater understanding of the other's needs, and they hold out hope for a long working relationship.

An Idaho rainbow trout.

If anything, anglers and irrigators alike believe that fewer mistakes will occur to harm the Henry's Fork if all of the players in the basin work together. And that, to everyone's delight, should make for a brighter future for irrigators and anglers in the Henry's Fork, and for the green drake, which is sure to emerge once again to stir the fancy of every angler on the river.

TANKS AND TALONS

Acutting, hot wind whips across the dry sagebrush plain just a stone's throw from a sheer basalt cliff overlooking the Snake River. The wind scours the ground, kicking up dirt from roads carved by military tanks and ash from fire-charred soil. A prairie falcon hovers in the breeze. Wings spread, it twirls and abruptly dives like a kamikaze pilot toward the ground, zeroing in on a Townsend ground squirrel. Thirty yards away, a .50-caliber machine gun tracer round plows into a stand of bluebunch wheatgrass and explodes. The squirrel's squeal is barely audible as the falcon seizes its prey in a cloud of dust and angles toward the cliff.

Steep basalt cliffs of the Birds of Prey Natural Area are home to one of the world's largest concentrations of hawks, eagles, falcons and other birds of prey.

Tank shells and prairie falcons might seem like an incongruous match, but for the last 40 years, the Idaho Army National Guard has trained here on the wind-swept plateau adjacent to the Snake River canyon. The tank firing range, called the Orchard Training Area, occupies about one-third of the 482,000-acre Snake River Birds of Prey Area, a national wildlife refuge. More than 600 pairs of a diverse mix of birds of prey nest in the cavernous basalt cliffs of the Snake River canyon. Many species, including golden eagles, prairie falcons and red-tailed hawks, hunt for prey on the plateau above. Officials estimate that

30–50 percent of the birds' prey base resides in the Orchard Training Area. Therein lies the rub. Can the National Guard conduct its exercises in a manner that does not jeopardize the nesting and hunting activities of the world's densest-known concentration of birds of prey?

That question raised a political ruckus in the late 1980s, when the Idaho Army National Guard proposed an ambitious modernization plan for its tank firing range. If approved, the range upgrade would make the Orchard Training Area one of two state-of-the-art tank ranges in the United States. The $13.6 million project would involve wiring a large area of the desert to set up 270 "hardened" targets, which would rise out of the ground at various speeds to challenge 290 charging tanks. Among the tanks used in the range are the M1 Abrams models, the type used in Operation Desert Storm in 1991. The modernized range would be available for other National Guard troops from throughout the nation, and that would mean a potential for increased impact on the raptors' prime hunting grounds. Morley Nelson, a world-renowned birds of prey expert from Boise, caught wind of the Guard's expansion plans and immediately launched a strenuous protest.

The Guard's modernization plan reflected an important change in American military policy since the Vietnam War. As the regular Army demobilized in the 1970s, U.S. defense

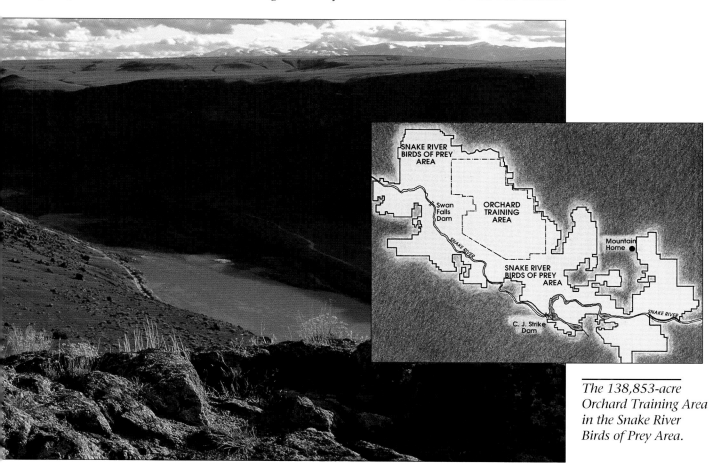

The 138,853-acre Orchard Training Area in the Snake River Birds of Prey Area.

179

At least 15 species of raptors nest in the steep walls of the Birds of Prey Area; above, golden eagle.

policy came to rely more on an "all-volunteer" force, principally Guard troops. That meant that Guard troops need to be able to train close to home and train with state-of-the-art equipment that would be used in real-life combat situations. Today, more than half the Army's combat forces are found in Army Reserve and National Guard units.

In 1988 the Guard went through a public process for approving the range modernization plan under the National Environmental Policy Act. The act requires an environmental impact statement for "significant" actions that pose a potential threat to the environment, and an opportunity for public comment and public meetings. The Guard's environmental impact statement, completed by a private contractor, CH2M Hill, did not address the issue of whether the Guard's stepped-up activities in the raptor refuge would pose an adverse cumulative impact on the area. Instead, it focused on the immediate details of what impact new buildings, training targets and spectator bleachers would have on 25 acres of the range. Nelson and others objected to what they perceived as a superficial analysis. More than 1,000 people wrote letters of protest, and Nelson declared he was ready to call in the national media if the Guard insisted on sidestepping the cumulative-impact issue.

After nearly a year of negotiations, the Bureau of Land Management worked out a compromise in the Birds of Prey Area: the agency authorized the Guard to move forward with its range modernization project, and Congress earmarked an additional $6 million for a battery of environmental studies. The BLM's decision notice concluded, "Because such a small area will be disturbed by construction, because the level of military training will not increase, and because 10 years of monitoring data indicates that raptor populations are relatively stable, we believe that the proposed projects can be constructed without harm to the wildlife values of the Snake River Birds of Prey Area."

As late as 1993 it was too early to draw any hard conclusions about the Guard's cumulative effect on the Birds of Prey Area. "We've got lots of data, but it's all preliminary," said Karen Steenhof, a raptor research biologist for the BLM. Five studies were being

Target practice — an M1 Abrams tank in the Orchard Training Area.

180

conducted to determine the Guard's impact on the birds and their prey: raptor distribution and use of the Orchard Training Area; hunting activities and range of prairie falcons and golden eagles using radio telemetry technology; raptor nesting densities and reproductive success, including whether raptors that nest close to the Orchard Training Area have more or fewer chicks than others; ground squirrel population surveys and reproduction; and use of vegetation and habitat by ground squirrels.

Pending the outcome of the studies, the Guard has dramatically adjusted its training activities to reduce the impact on the birds of prey. In 1989 the Guard hired Marjorie Blew, a professional wildlife biologist, to beef up its environmental commitment in the Orchard Training Area. Since then, Blew has set up a number of safeguards to minimize the Guard's impact on the range, including environmental-awareness training for incoming Guard troops, the use of Guard troops to plant native shrubs in areas where wildfires have wiped them out, the closure of habitat areas to Guard training and the establishment of a permanent fire crew that accompanies Guard troops during training missions.

A helicopter equipped with a 3,000-gallon water bucket is on standby in the event of wildfires. The fire crew has drastically reduced the amount of natural habitat scorched by wildfires, from an average of 10,000 acres per year to less than 135 acres in 1992, Idaho's worst-ever drought year.

All of these measures and more led to Blew's receiving a national natural resources conservation award from the U.S. secretary of defense in April 1993. In addition, the Idaho Army National Guard received a similar award from the Department of the Army in 1992, the first time that the Guard had ever received such an honor. Blew's leadership at the Orchard Training Area has led to her being named a national coordinator for military training-range conservation efforts.

Ferruginous hawk watches its young feed.

"Managing natural resources on land with less than 8 inches of annual rainfall is a difficult assignment in itself, but when the land is also home to intensive military training with heavy tanks and supporting equipment, the task is truly monumental," the Guard's award-nomination entry said. "Picture the task of accommodating the training needs of highly maneuverable armored vehicles, providing forage for raptors, many of which are endangered or of special concern, protecting endangered plants, ensuring protection of archaeological sites, and taking care of wildlife in general ... all on the land of another agency which heavily grazes cattle and sheep. Yes, picture that task, and you can begin to see the task that Marjorie Blew faces at the Orchard Training Area."

As secretary of the interior, former and future Idaho governor Cecil Andrus protected the raptor refuge under an administrative withdrawal in 1980. Such withdrawals expire after 20 years. In the 1992 session of Congress, Rep. Larry LaRocco, D-Idaho, drafted legislation that would permanently protect the

BLM biologists study nesting golden eagles.

refuge, but Sen. Larry Craig, a Republican who is a staunch defender of the livestock industry, blocked the bill due to concerns about how local ranchers and livestock grazing might be affected. At the year's end, Craig and LaRocco reached a compromise, and Congress approved permanent protection for the Birds of Prey Area in 1993.

Echoing ranchers' concerns, the Idaho Farm Bureau has objected to "the increasing tendency of our federal land management agencies toward managing vast areas of our public domain for the maximum protection of a very narrow spectrum of outputs." The existing management scheme for the raptor refuge, as outlined by a 1985 BLM management plan, puts the highest priority on protecting raptor habitat. The plan has four goals: (1) maintain the nesting raptor populations at or above the minimum populations determined in the period from 1975 to 1981; (2) provide for other, compatible uses; (3) coordinate research to support management needs; and (4) publicize the results of that research and information on management activities.

Once the research projects conclude, certain Guard activities or other uses such as livestock grazing may be adjusted to protect the refuge's integrity. It is doubtful, however, that a need will arise to exclude the other multiple-use activities.

The dispute over the Birds of Prey Area reminds us that land conflicts between competing uses also can be a contest over values. Few would question the need to protect the marvelous raptors, but national defense, for many Americans, is a fundamental value too. Certainly the training could be moved to another location, but there would be a cost; and the rocketing expense of government is an equally burning concern. At this point, it appears that the Guard, the BLM and raptor advocates such as Nelson have found the middle ground where most uses can continue to thrive in the refuge in relative harmony. It is interesting to note that scientific research — that is, to answer the question of whether the tanks are really harming the birds of prey — became the key to resolving this conflict. This is somewhat of an anomaly in natural resource debates, because science seldom plays such a decisive role in instances of severe value conflicts. As long as Congress

Gov. Cecil D. Andrus. During the closing days of the Carter administration, Andrus, then secretary of the interior, extended the moratorium on development in the Birds of Prey Natural Area; below, Guard soldiers plant sagebrush in a field damaged by fire.

leaves the BLM some management flexibility in the future, the Birds of Prey Area example suggests that any new concerns that arise might be addressed in a cooperative fashion.

While the Army was learning to live with the raptors, just 60 miles to the south of the Birds of Prey Area, a much more fiery land-use conflict has arisen, this time over the proposed expansion of an air combat training range in the heart of scenic canyons in the Owyhee River plateau. Here, competing interests have been mired in gridlock for several years — the values in conflict are entrenched and the issues are far from being resolved. In this dispute, Gov. Cecil Andrus, who is considered by many to be the state's leading elected environmentalist, has supported the efforts of Mountain Home Air Force Base to create a new training range in the Owyhees. At one time, Pentagon planners saw the Owyhees as a blank spot on the map, a place where cows and wild animals are far more plentiful than people. But the Owyhees, though still scarcely populated, contain a network of unique canyons that are renowned for their natural beauty. Moreover, Shoshone-Paiute Indians fear the supersonic flight path will disrupt sacred burial grounds and the tranquil life of the Duck Valley Indian Reservation on the Idaho-Nevada

An F-15 flies above the Idaho desert; below, a cement-filled (BDU 33) bomb dummy unit.

border. A variety of conservation groups organized as the Owyhee Canyonlands Coalition have expressed concern over the possible impact of the range on wildlife. A 1993 Idaho Department of Fish and Game report warned that the proposed range has an "extremely high likelihood" of disturbing 1,700 pronghorn antelope, the largest herd in the state, as well as the rare bighorn sheep that live in the canyons. Flares and practice bombs may start grass fires, according to the report.

The range proponents have run into opposition. But from the standpoint of national defense and the Idaho economy, the stakes are high. Mountain Home Air Force Base means $1 billion in payroll and spinoff revenues to the state of Idaho, Andrus has claimed. The base is also Elmore County's leading employer. Although there are no immediate plans to close the base, Andrus has long maintained that a new or expanded range will protect Mountain Home from federal budget cuts. Backers of the base argue that southwest Idaho is one of the finest places to fly and train in the nation. Base backers doubt that the "startle effect" of the thundering jet will drive antelope out of the region. And the Air Force believes that blazes started by flares can be quickly extinguished, limiting the fire damage. Even while the issue remains undecided the Air Force made the decision to move a so-called "composite wing" to Mountain Home AFB. The wing is a new nationwide concept in which a variety of aircraft would be based at one facility for coordinated training in a more realistic battle setting.

The heated debate over the proposed high desert training range illustrates how Idahoans of all viewpoints hold entrenched and deep-seated beliefs about the values of wide open spaces. Still, there may yet be room for a compromise as state and federal officials work to accommodate rival positions on the management of the desert and highest use of the land.

THE IDAHO NATIONAL ENGINEERING LABORATORY

How times have changed at the Idaho National Engineering Laboratory since the nearby town of Arco introduced the world to nuclear-powered electricity in 1955. It was an exciting moment for nuclear engineers, one that marked the dawning of the age of nuclear power in the United States. Today, the INEL, located about 60 miles west of Idaho Falls on a mile-high sagebrush-dotted flat, is an international nuclear research center, a temporary storage depot for federally-owned commercial and military nuclear waste, a nuclear submarine and surface ship training facility for sailors in the U.S. Navy and a designated Superfund toxic waste site. It is also Idaho's largest employer with an annual budget exceeding $1 billion.

In more than 40 years of operation, the INEL has evolved from the world's largest nuclear research facility for the U.S. departments of Defense and Energy, a vital link in the chain of nuclear weapons plants around the nation, to one that is searching today for more peacetime scientific research and innovative ways to reuse, condense or dispose of all kinds of nuclear waste. INEL researchers also hope to demonstrate a new type of nuclear power reactor for public use that would use recycled uranium for fuel and generate far less waste byproducts.

For Idaho citizens, the biggest issue at the INEL is nuclear waste — the environmental legacy of storing radioactive and toxic waste byproducts above the Snake River Aquifer and the haphazard dumping of other plutonium-contaminated wastes at an 88-acre site inside the 890-square-mile compound. The INEL stopped dumping waste in 1969, but residents in the Twin Falls area fear that waste byproducts may eventually migrate 150 miles to the southwest and foul seemingly abundant ground-water that is used for drinking and crop irrigation. So far, research shows that only a small percentage of waste has migrated beyond INEL's boundaries, and that portion has not gone beyond an area immediately adjacent to the INEL.

Meanwhile, Democratic governor Cecil Andrus scored political points with some Idaho citizens by taking a hard line against INEL being used as a permanent storage site for high- and low-level nuclear waste from other nuclear power plants in the nation. As it is now, INEL temporarily stores 132 tons of nuclear waste from the Three Mile Island incident, as well as spent fuel cores from naval submarines and its own waste. Andrus has made it clear that Idaho will not allow any more commercial nuclear waste to be stored at INEL, putting pressure on the Department of Energy to open a permanent nuclear waste depository, a prospect that was still 20 years from approval in the early 1990s.

Heavy metal hatches at INEL's Radioactive Waste Management Complex; above, a truck hauls waste.

The Idaho cowboy, the backbone of a $600 million industry, roams 16.5 million acres of high desert rangeland.

MARLBORO MAN IN TRANSITION

Ah, the mythical romance of the West — nothing seems to evoke those feelings of the big wide-open valleys, snowcapped peaks and grazing livestock more than the cowboy, the Marlboro man. How could one imagine from a glance at those nostalgic pictures of cowboys tending a fire in the afterglow of a mountain sunset that these down-home folks are under attack like never before? How could it be that some Americans are turning on their favorite western heros?

In his 1992 book, *Crossing the Next Meridian — Land, Water, and the Future of the West*, University of Colorado law professor Charles F. Wilkinson describes western cattlemen as the "Lords of Yesterday," a group of hardworking, independent people who operate under 19th-century codes in today's world. Events in the last two decades, however, have threatened the ranchers' traditional way of life, particularly their ability to exercise free rein on public rangelands. One of the most publicized incidents to illustrate that point occurred near Twin Falls in the hot summer of 1990.

Twin Falls District ranger Don Oman, a soft-spoken 26-year veteran of the U.S. Forest Service, oversaw 320,000 acres of the Sawtooth National Forest, south of Twin Falls, an area called the South Hills. When he came to the district in the late 1980s, he was "appalled" by the condition of the forest lands, particularly the fragile ribbons of green vegetation along small streams. He blamed the ranchers for overgrazing the land and his predecessors for failing to enforce the law protecting riparian areas. "It's worse than any place I've seen," he remarked in 1990.

Oman told ranchers they could not keep their cows on the range any longer than their federal permits allow; he told them to keep their cows out of the creek bottoms; he told them to repair their fences and water developments. He did not get much cooperation. Though it may be hard for urban citizens to understand, western ranchers do not take kindly to getting ordered around by "the feds"; they usually can call their local congressman to attempt to put pressure on officials who might have other ideas about the way they run their livestock on federal lands. "There's an old saying in the Forest Service that if you cross a stockman, you can expect to be shipped to Siberia tomorrow," Jim Prunty, a retired Forest Service official, told the *New York Times*. "Don Oman was the first person ever to say no to these people and that's why he's in trouble."

To Oakley rancher Winslow Whiteley, Oman went too far when he brought a group of armed federal agents to the South Hills, confronted a group of cowboys and told them that they were violating the terms of Whiteley's grazing permit. Whiteley was furious and immediately called for Oman's transfer. "Either Oman is gone or he's going to have an accident," he told the *Times*. "Myself and every other one of the permit holders would cut his throat if we could get him alone." Asked if he was, indeed, threatening Oman's life, Whiteley said "yes," adding, "If they don't move him out of this district, we will."

Whiteley called then-senator Steve Symms, R-Idaho, a staunch conservative and an outspoken critic of the Forest Service in Congress, who met with agency officials and arranged Oman's transfer. But Oman had anticipated the maneuver and filed a whistle-blower complaint with the Forest Service's inspector general. He effectively blocked the action by the agency, Symms and Whiteley.

Oman remained on the job while awaiting a transfer, but the show-down with Whiteley led to national stories in the *New York Times*, *People*, *Sierra* and *Audubon*. Those stories brought the public land grazing issue to

Idaho Cattleman's Association, 1973.

IDAHO CATTLEMENS ASSOCIATION

"BEEF – FOOD FOR THE WORLD"

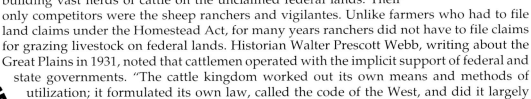

the nation's attention, much to the delight of environmentalists. The following year, the National Audubon Society ran an hour-long special on public television entitled the "New Range Wars," outlining the environmental threats of overgrazing and the political vise-grip that ranchers have had on public agencies. The show spotlighted the federal government's $38-million-a-year predator control program, in which federal trappers kill thousands of coyotes and lesser numbers of black bears, mountain lions and foxes to prevent the loss of cattle and sheep on public lands. The National Humane Society wants to end the predator control program, and some environmentalists have called for the removal of cows from public lands. The radical environmental group Earth First! crafted the slogan, "Cattle Free by '93." Livestock grazing on public lands remains a contentious issue, and will be for many years to come.

To understand the ranchers' point of view in this debate, one must return to the genesis of ranching in the West. After the Civil War, the great cattle drives into the remote interior of the West began, emanating from Texas and Mexico. According to author Charles Wilkinson, some 600,000 Texas longhorns were driven north and west out of Texas in 1871. By the mid-1880s, the number of livestock in Nevada, Oregon, Montana and Idaho had skyrocketed to some 26 million cattle and about 20 million sheep. Under the Homestead Act of 1862, ranchers could secure a base of 160 acres adjacent to public grazing lands. This arrangement usually consisted of homesteading on the edge of a mountain valley, with a dependable water source, a place to grow hay and pastures for the livestock. In the late 1800s and early 1900s, cattlemen built large empires by building vast herds of cattle on the unclaimed federal lands. Their only competitors were the sheep ranchers and vigilantes. Unlike farmers who had to file land claims under the Homestead Act, for many years ranchers did not have to file claims for grazing livestock on federal lands. Historian Walter Prescott Webb, writing about the Great Plains in 1931, noted that cattlemen operated with the implicit support of federal and state governments. "The cattle kingdom worked out its own means and methods of utilization; it formulated its own law, called the code of the West, and did it largely upon extra-legal grounds," he said.

In 1906, at the urging of President Theodore Roosevelt and the U.S. Forest Service's first chief, Gifford Pinchot, the agency began charging a fee of 5 cents per animal unit month (AUM) — the equivalent of five sheep or one cow-calf pair grazing for one month — for grazing on national forest lands. But the nonforested federal lands still had no rules or fees for grazing. These lands, now controlled by the Bureau of Land Management (BLM), were called "the lands nobody wanted" — vast amounts of dry, mountain territory that homesteaders passed over for base farms or ranches. In 1934, on the heels of a prolonged drought in the West and the Dust Bowl years, Congress passed the Taylor Grazing Act, which created the National Grazing Service. The service charged ranchers 5 cents per AUM for the privilege of grazing on federal lands, but some 50 years of essentially free use of those lands had left the indelible impression that ranchers "owned" them, an impression that still exists to a certain extent today. Congress created the BLM in

Anticattle logo, 1986

Cattle have grazed southern Idaho since commercial ranching began during the Civil War. Today about 2,000 Idaho ranchers have federal permits to graze more than 200,000 cattle, but political pressure and the threat of erosion and desertification have forced the U.S. Bureau of Land Management to restrict open-range cattle ranching; above, a cow in the wind-blown Snake River country; left, Spanish longhorns reached Idaho via Texas and California, from a watercolor by William H. Jackson.

TOOLS OF THE TRADE

A shovel and rubber boots were symbols of prosperity and hope to early settlers on the Snake River Plain. Today, a shovel mounted on a trailbike or pickup is the key that will open any conversation or provide a stool in any coffee shop on the plain.

The technology of farming may have changed, but the dedication is just as intense as it ever was. Some say dairy farmers are the most dedicated people in the world, never failing to meet the milking schedule. And there is the potato/beet grower who excuses himself from the Sunday table, dons his boots and rattles off down the dusty road in his pickup to shift the water. Or, there is the farmer who bolts upright in bed when he hears his water pump shut on his pivot a mile away.

There are few in this world who are more dedicated or live closer to financial destruction than the agriculturalist whose tool of the trade is the irrigator's shovel. They explode from their beds by 5 a.m., shift the water, hit the coffee shop by 9 a.m., and by 10 a.m., the quiltlike pattern of roads on the Snake River Plain are stitched with lint-balls of dust thrown by the pickups as they charge off in all directions — man, shovel and dog, all working to make what was once a harsh landscape yield yet another crop.

Southern Idaho ranch

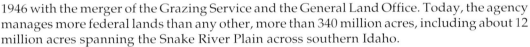

Overgrazed BLM land on Paiute Creek, Owyhee County.

1946 with the merger of the Grazing Service and the General Land Office. Today, the agency manages more federal lands than any other, more than 340 million acres, including about 12 million acres spanning the Snake River Plain across southern Idaho.

As Wilkinson pointed out in *Crossing the Next Meridian*, the BLM has always been underfunded and been denied the political backing to do its job: enforcing sustainable stewardship on its lands. In 1946, for instance, the Grazing Service tried to boost grazing fees to fair market value, and western congressmen slashed the agency's budget in half for trying to meddle with the ranchers' way of life. Indeed, scholars such as Phil Foss pointed out that the BLM was a "captive" agency, captured by the very interests it was supposed to regulate.

The BLM, the ranchers and western congressmen formed a tight private/government "iron triangle" that dominated public land policy in the West up to the mid-1970s. It worked like this: Congress appropriated grazing fees and other funds for livestock improvements; local grazing boards divided up the funds at the local BLM offices; and the ranchers installed the improvements on the ground, effectively adding value to their ranching empires. Many BLM employees came from small western ranching towns; the congressmen who oversaw grazing policy were from the intermountain states such as Nevada, Wyoming and Idaho. Thus ranchers had instant access to both Congress and the agency. This escaped public scrutiny because few others cared about BLM grazing policy.

Although most ranchers point to the Taylor Grazing Act as the turning point in range stewardship in the West, environmentalists argue that it was not until the 1970s — when environmentalism first emerged as a budding political force — that Congress gave the BLM the legal mandate to manage its lands for multiple uses such as wildlife and recreation. The law, the Federal Land Policy and Management Act (FLPMA) of 1976, was far-reaching at the time, and since then the public at large has been battling to influence management of a bigger share of BLM lands. The enactment of FLPMA later ignited the Sagebrush Rebellion — a crusade to turn over the BLM lands to the states or private ownership. There were two reasons for the Sagebrush Rebellion: (1) FLPMA strengthened the BLM's ability to regulate public land use, and particularly that used by ranchers; (2) The new law also made it clear that the BLM lands would remain in public ownership. Initially, the Sagebrush

Rebellion was given great hope by the election of President Ronald Reagan in 1980. Reagan responded by appointing archconservative attorney James Watt as interior secretary. Watt's positions in support of the Sagebrush Rebellion and the perceived exploitation of public lands raised a fury among some of the American public. In addition, the states realized that they could not afford to manage all the federal public lands, and if the federal lands were sold to the highest bidder, ranchers realized that they might be outbid by large corporate interests, foreign and domestic.

Today's controversy over the range focuses on two issues:

❑ Grazing fees — whether ranchers should pay what their critics call "market value" for the privilege of grazing livestock on federal lands, or whether they should be charged an incentive fee on the basis of land stewardship: "good" ranchers would continue to pay reduced fees for grazing rights, and "bad" ranchers would have to pay fair market value.

❑ Overgrazing — cows loafing in creeks and damaging the vegetation or riparian areas along the stream banks. Research has shown that these plants and trees are critical habitat for nesting birds, fish and aquatic insects. BLM and Forest Service studies indicate that more than half of the public range in 13 western states is in fair to poor condition. The future trend will likely focus on ranchers improving livestock management to prevent cows from concentrating near stream bottoms for weeks or months at a time.

Ranchers argue that increased grazing fees will put many of them out of business, and that if they are forced out, then the government should buy them out at fair market value. This would not be an unprecedented action for the federal government to take; it has done so many times to preserve critical habitat for parks and other lands. Environmentalists argue that if Congress increases grazing fees, ranchers will run fewer cattle on public lands and that the damage to streams and riparian areas will therefore decrease. Interestingly enough, a Democratic Oklahoma rancher, Congressman Mike Synar, has led the charge in Congress to increase the fee. He says it is unfair for western cattlemen, who raise less than 3 percent of the nation's beef, to receive a subsidy for grazing on public lands, when eastern ranchers get no breaks for grazing on private lands. But National Cattle Association officials have consistently supported the western ranchers, endorsing the view that grazing livestock on federal lands involves a number of extra hidden costs that justify the lower fee.

Meanwhile, a new generation of ranchers has taken up the task of improving land stewardship. The Little family in Emmett, for instance, adopted an innovative system of rest-rotation in the late 1960s, well before many had heard of the concept that revolves around the

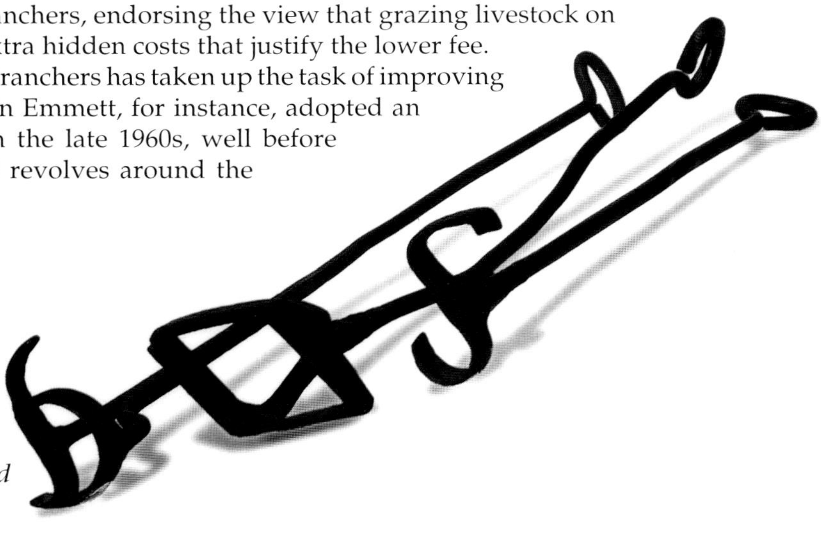

Three Idaho brands: the diamond, the 76 connected and the Lazy A.

193

Disputes over grazing fees, water rights and endangered species are threats to the livestock rancher's traditional way of life; above, herding cattle to a branding corral, Owyhee County; right, Bruneau rancher Gene Davis worries that an endangered species listing for snails will hurt his large cattle operation.

principle of resting fenced grazing pastures for at least one year to allow them to recover and rebound with thrifty grass. David Little, a former state senator and chairman of the Legislature's Natural Resources Committee, has two sons, Brad and Jim, who run cattle and sheep operations on public lands in southwest Idaho. Both Brad and Jim Little have been active in congressional debates on grazing fees, and Brad has spent countless hours on predator control issues and the question of whether to reintroduce Rocky Mountain gray wolves to Yellowstone National Park. In recent years Brad has attended several courses taught by Allan Savory, a Rhodesian native who endorses the concept of holistic resource management. The Littles' efforts toward good land stewardship have gone a long way toward protecting their federal grazing allotments from criticism by environmentalists, and they have found that they can put more pounds on their cows with higher-quality grass.

On the Camas Prairie, just north of the Bennett Hills near Gooding, many ranchers have embraced the use of beaver to restore badly degraded riparian areas. Lew Pence, a U.S. Soil Conservation Service official, believes that beavers should be returned to many mountain streams to help rebuild the down-cut creeks. Ranchers who have reintroduced the beaver are impressed with the results. Beaver dams raise the water table of a creek, increase the amount of grass in the meadows surrounding the beaver ponds and, in so doing, create more habitat for birds and wildlife. "I've seen more land inundated with water than I had before, but the overall effect of

Rodeos showcase the skills of the ranch hand as the West's own spectator sport. More than 8,000 professional cowboys compete on the rodeo circuit; right, rodeo competition in Jordan Valley; below, an Idaho Cowboy Association championship bull-riding buckle.

SNAKE RIVER SHO-BANS

Sho-Bans in traditional dress at a 1991 pow wow.

Modern times have been hard on Shoshone, Bannock and Paiute peoples — former nomads of the sagebrush desert who retain lands on the Fort Hall Reservation near Pocatello and the Duck Valley Reservation on the Idaho-Nevada line. Jobs are tough to get. Many of those who cannot find positions in tribal government or on ranches and farms live in poverty. Reservation officials estimate that more than half the Indian children drop out of school.

Clearly, even in the 1990s, integration into white society has been a grim process, and many tribal members still resist it.

But there's one element of the white culture that brings new hope: gambling.

In the late 1980s, Congress passed a law that invited tribes to negotiate with states to build casinos and bingo

raising the water table has so many benefits, it far exceeds the value of the land lost," says Fairfield rancher Jon Mellen in *High Country News*. So far, Pence has aided ranchers in planting about 100 beaver in 25 creeks in southern Idaho.

What is most convincing to Mellen is the increased productivity of his land. Using 6,000 acres of private ground and 4,000 acres of state and BLM land, he has divided up the ranch into 22 pastures with New Zealand solar electric fence. "I've doubled the productivity of the land with the New Zealand fence and the beaver," Mellen says. "We've averaged two pounds per day per head with the rotation. But if you just turned the cows loose, you might get only 1.5 pounds per day per head. Of course, not everyone has the quality of pasture land that we do."

Mellen, the Littles and others are not alone in trying to manage their ranches in a more progressive fashion. Similar trends are occurring in Oregon, Colorado and Montana. TV tycoon Ted Turner and his wife Jane Fonda went so far as to sell off the cattle on their Montana ranch and turned over the pastures to buffalo. Other ranchers are trying to make a living with game farming, a European concept in which landowners charge hunters thousands of dollars (on top of license and tag fees) for the opportunity to shoot a big bull elk on private land. Still others are supplementing income with dude ranching, offering horse rides and weekend getaways for urbanites who want to soak up a piece of the country.

halls on their reservations to raise much-needed income.

At Fort Hall, the Shoshone Bannocks have built a fancy new bingo hall that employs about 80 people and provides a new source of revenue for the tribes. "All of the people working for us were unemployed before," said Nathan Small, Sho-Ban gaming manager. "A lot of these people have been taken off the welfare rolls. We're proud of that. They want to earn their own living; they want to work."

There is no bingo hall at Duck Valley, although the Shoshone-Paiute tribal council is looking at proposals.

Hazardous waste shipments through the Fort Hall Indian Reservation have kept Jeanette Wolfley, general counsel to the Shoshone-Bannock Tribes, busy in court; below, Fort Hall bingo sign.

Efforts to go a step further and build a casino at Fort Hall, Duck Valley or elsewhere were set back when the state legislature in a special session in the summer of 1992 proposed a constitutional amendment to restrict Indian gaming. Voters adopted the amendment in the November 1992 election.

Undaunted, Native Americans have turned to the courts to actively protect their lifestyle and heritage. In 1990, Fort Hall Sho-Bans negotiated a water rights settlement with the state of Idaho to reserve more that 500,000 acre-feet of water in the Upper Snake River for the tribal farms, fish and wildlife. As part of the same Snake River adjudication process, the Shoshone-Paiute Tribes have been negotiating for creek water impounded by Idaho and Nevada farmers above the Duck Valley Reservation. Tribal governments also are working on reserving in-stream flows for salmon and steelhead in the Salmon River Basin.

Today, these public rangelands in the Snake River Plain are no longer the lands nobody wants. Increasingly, they are lands that more people want to use. Though it is difficult for ranchers to accept change, it appears that their ability to improve land stewardship and cooperate with public land managers and recreationists will largely determine their fate. Indeed, the winds of political change are swirling over the public rangelands, but at the same time ranchers clearly hold the key to their future. If the traditionalists persist in defending the ways of the 19th century, ranchers could stand to lose it all. If more progressive approaches prevail, then perhaps ranchers can find a new sense of balance and a sustainable future. Says Wilkinson, "Ranchers have contributed much to the West, and they deservedly hold a prime place among those with access to the public rangeland, but they have no right to a monopoly or to practices that injure the ground. If we look first to the needs of these long-neglected lands, the western range will provide benefits aplenty."

Flooded by cars and almost 3 million tourists each year, Yellowstone, a national treasure, is a victim of its own success; right, a Union Pacific travel brochure, 1913.

YELLOWSTONE — A MANAGEMENT MODEL

Many people know that Yellowstone is the world's first national park. The national park idea took root here and has since spread throughout the world. Thus Yellowstone is of tremendous symbolic importance to many people, and what happens to this park is always very closely observed.

Yellowstone has often served as a laboratory for ideas about our national parks and how they should be managed. Not surprisingly, these ideas are often the source of much conflict.

Visitors to Yellowstone may wonder about its curious shape, which is basically a rectangle. The park's boundaries were drawn in 1872, and they reflect congressional attention to the protection of the most spectacular resources of the area, such as the geysers. When Congress created Yellowstone National Park, it was thought that these square boundaries would do adequate justice to whatever additional natural wonders might be discovered within them, but the boundaries did not appear to reflect concern for the biological resources of the park, such as the grizzly bear. This lack of concern is understandable from the perspective of 1872, because the need to protect biological resources was simply not a pressing concern at that time. Today, however, there is a great deal of concern about the park's boundaries and whether they are adequate for all of its natural resources.

Yellowstone, and the area around the park, has come to be known as the Greater Yellowstone Ecosystem. Many individuals, interest groups and government agencies have begun to discuss how to manage the entire ecosystem in a more consistent and comprehensive manner. This discussion does not focus on enlarging the park. Instead, talk about more consistent management tends to revolve around better protection for the natural resources of the system. Why isn't the area already managed that way, one might ask? The answer: conflicting values.

Much of the land around Yellowstone is under the jurisdiction of the Forest Service. This agency has been given contradictory directions by Congress on how to manage land under its jurisdiction. The Forest Service manages lands for "multiple use," which means that timber harvesting, geothermal exploration and mining occur on national forest lands, often not far from nearby national parks and wilderness areas. Historically, in the intermountain West, the extraction of natural resources has had at least as much importance as their protection. Understandably, the agency in charge of managing public land for resource extraction will likely be predisposed to allow some of that extraction to occur. There are also constituent groups that support multiple-use management on these public lands and would thus resist an "ecology-first" management style out of fear that it would reduce natural resource extraction activities. There are also private lands scattered throughout the area, on which activities may occur that have the potential to affect the natural resources of the area.

WHERE GUSH THE
GEYSERS

 UNION PACIFIC RAILROAD
OREGON SHORT LINE RAILROAD
THE NEW AND DIRECT ROUTE TO
YELLOWSTONE NATIONAL PARK

The great Yellowstone fire of 1988 rekindled a heated debate over the so-called "let burn" policy in America's national forests; right, the inferno near Old Faithful Lodge; inset, fighting the fire at Tower Junction.

Photographers line up for a Yellowstone grizzly at Dunraven Pass; below, feeding the bears, 1913.

Perhaps the key event that focused attention on the area as an ecosystem was the decline of the grizzly bear population. Scientists, environmentalists and others came to realize that the bear, though protected in the parks of the area, was in need of greater protection outside the park boundaries. Although Yellowstone protects some 2.2 million acres of bear habitat, the animal actually appears to use and "need" more than 5 million acres. Many of these other acres are on U.S. Forest Service lands.

Concern for the bear led to formation of the Interagency Grizzly Bear Committee, made up of representatives of all the area's public land agencies. This coordinating committee has created various land management designations based on the bear's habitat, not jurisdictional boundaries. Other interagency committees in the Yellowstone area have followed, each concentrating on other species.

The concern for the bear has thus led to a more general concern with the entire Greater Yellowstone Ecosystem, and the need to manage it in a more consistent manner. To that end, in 1987 all of the Yellowstone area national forest and national park management plans were put together, for the first time, in one document. The Aggregation, as it came to be called, did not mandate any uniform or systemwide management for the area. It did, however, provide the first comprehensive look at all the management activities, both ongoing and proposed, for the area.

By the late 1980s, however, events became increasingly contentious. The National Park Service and the Forest Service decided to move more rapidly toward integrated management of the Yellowstone area. The two agencies issued a draft document titled "Vision for the Future: A Framework for Coordination in the Greater Yellowstone Area." That document called for a management approach in the region that would do two things: (1) conserve naturalness and ecosystem integrity, and (2) encourage biologically and economically sustainable opportunities.

What happened after the issuance of the "Vision" document amounted to a political firestorm. Many local people, suspicious that the document implied a drastic curtailment of traditional multiple-use activities, convinced regional elected officials and the Bush administration to put a halt to the implementation of the Vision process.

How should we resolve conflicts of this kind in the future? Many sound scientific arguments can be advanced for managing areas like Yellowstone in a more coordinated and systematic way. But, in order to achieve such a change, we need to have some sort of general consensus or agreement that such a change is in the public interest, and thus "good." It is clear that the Park Service and the Forest Service may have assumed that consensus existed, when, in fact, it did not. One thing is certain, however: the battle over the management of the Yellowstone area has just begun.

The problem of coordinating activities in and around Yellowstone can also be seen from another perspective. It is becoming increasingly apparent that our national parks

are being impacted by activities outside the boundaries of the parks: Air pollution obscures the visibility of many national parks such as the Grand Canyon; land use conflicts have developed near the boundaries of many of the parks; and the water quality of many parks, such as the Everglades, is threatened. The problems are often difficult to resolve because, once again, opposing values compete with one another.

At present, the National Park Service does not have authority to do anything about adjacent lands. Congress in recent years has made several attempts to give the agency more authority, but these attempts have failed, primarily because many people do not want the protection of national parks to be given a higher priority than economic development outside of the parks.

It is also difficult to manage all of the units of our national park system in the same way. Some are large natural areas like Yellowstone, where the ecosystem concept may make sense. Other units are historic sites, like Gettysburg, where reliance on the land planning process of local government may be the most effective approach to solving problems.

Yet Yellowstone remains a special place. It will always be the first national park, and its protection, however it is accomplished, should remain a high priority. But to protect it requires us to give it a higher value than many other parks in our political system. As the

A NATIONAL PARK IN IDAHO?

Although Idaho sits in the middle of the spectacular Rocky Mountains, the state has no national park within its boundaries. Idaho, however, is not lacking in areas that both qualify and have been proposed as national parks.

Once an area meets the National Park Service criteria, there must be a great deal of support within a state before the park can be created. Members of Congress must approve the designation, and no significant group can be in opposition. These conditions have never come to pass in Idaho for a variety of reasons:

❑ Since a national park would not allow hunting, there has always been opposition from Idaho hunters and the state Department of Fish and Game.

❑ Two past candidates for park status, the Sawtooth Mountains and Hells Canyon, are administered by the Forest Service as national recreation areas. The Forest Service has done an excellent job convincing Idahoans that these areas fare well under that agency's management.

❑ Fearing the crowds and overdevelopment of "another Yellowstone," Idaho environmental groups have, as a rule, opposed national parks.

There is one area, however, that does appear a possibility to receive national park status — the Craters of the Moon. According to opinion surveys, a majority of Idahoans would like to see the Craters became a national park. An enlarged Craters would clearly meet National Park Service standards. Today, there is a quiet momentum, sometimes behind the scenes, to build support for a Craters of the Moon National Park. Perhaps soon that momentum will reach a critical mass and Idaho will have its first national park.

Sawtooth Mountains near Stanley; left, Hells Canyon.

Yellowstone after the 1988 fire; below, park ranger with bull elk skull at Gibbon Meadows, Yellowstone.

superintendent of Yellowstone has said: "We've got to go beyond the glittering generality that we should 'maintain the integrity of the ecosystem.' That's like saying 'I'm for wise use.' There have to be sovereign conditions in the ecosystem that we can all agree to that go beyond our agency missions and mandates in order to ensure this integrity." It is the development of those sovereign conditions that is the problem. New legislation mandating the protection of the ecosystem is likely to be resisted because it gives highest priority to protection. But most types of coordination, or cooperation, have problems as well. As political scientist Harold Seidman has pointed out: "When conflicts result from clashes in statutory missions or differences in legislative mandates, they cannot be reconciled through the magic of coordination. Too often organic disease is mistakenly diagnosed as a simple case of inadequate coordination." He went on to say: "If agencies are to work together harmoniously, they must share at least some community of interests and compatible objectives; problems cannot be resolved by coordination."

Again, the struggle over the Yellowstone area is a clash of different values — the question of which public values are the most important. Ultimately, it may be individual species that are protected, rather than any specific area of land. What is clear is that some of the most insightful thinking, planning and activity in the realm of parks and public policy is occurring in the name of Yellowstone and its surrounding area.

Tourism and recreation are changing the political landscape of a largely agricultural state; left, rafters on the Bruneau River.

A SUSTAINABLE FUTURE

Many decisions about natural resources frequently are stalled in political gridlock. The remarkable cooperation between the National Guard, raptor advocates and the BLM in the Snake River Birds of Prey Area is nowhere evident in the debate over a new Air Force training range just 60 miles to the south in Owyhee County. Ecosystem management in the Greater Yellowstone area seems light years away from a holistic management system for the Snake River from its headwaters in the south end of Yellowstone to its mouth in eastern Washington. Yet, such a solution would be considered watershed management, a concept pioneered well over 100 years ago by western pioneer John Wesley Powell. This concept of managing lands by watersheds rather than artificial political boundaries is especially applicable to the Snake River because it flows nearly from source to mouth in one state.

A clear consensus on the protection of the world-famous Henry's Fork also does not mean it will be easy to find a way to save the Snake River salmon runs or endangered snails. As it is now, each resource issue pits one interest against another, tempers flare and, in the end, the players depart with precious little understanding of the opposing points of view.

Why can't citizens view these individual debates over the future of the Snake River Plain as the potential groundwork for a much larger debate about the future? For clearly, the plain is at a crossroads, where some resources, such as salmon, snails and the Snake River itself, have reached the breaking point.

A trout farm in the Hagerman Valley.

TROUT FARMING

Trout are to Idaho as, well, potatoes are to Idaho. More than 75 percent of all trout consumed in the United States are grown in Idaho. Although people everywhere readily picture a trout dinner as originating in the cold, rushing streams of the Idaho high country, most of the 40 to 50 million pounds of Idaho trout are raised in hatcheries along the Snake River between Twin Falls and Bliss.

Idaho trout farming began in 1928 when the Snake River Trout Company built a hatchery in the Magic Valley near Thousand Springs. The water that feeds these springs originates in the mountains of central Idaho and moves southward until it reaches the Snake River Plain. Absorbed by the rock, the water travels more than 100 miles underground before it gushes into the Snake River at a rate of up to 2 million gallons per minute. The many miles of travel through this natural rock filter leaves the water crystal clear, at a constant temperature of 58 degrees — ideal for growing rainbow trout. Today more than 120 Idaho hatcheries annually produce a trout bonanza worth $70 million.

Front Street, Boise (1994) by Peter Wollheim.

The question is: Can we come together as a community on the Snake River Plain and plan a collective future while we still have the necessary resources to sustain us? Are we up to the task? In *Crossing the Next Meridian*, Wilkinson puts it this way, "Western communities can either take charge of the future by adopting some form of conscious management and direction, based on full and brightly etched visions of the future, and sustain the West's lands, waters and way of life; or western communities can continue to abdicate — by allowing developers to charge ahead with few restraints — and surrender the distinctive qualities of the West within a very few decades."

The late Wallace Stegner, in his final book *Where the Bluebird Sings to the Lemonade Springs — Living and Writing in the West*, reminds us that the task of conserving our natural resources will take constant vigilance and work.

> Considering the mood in which the continent was settled, and the amount we had to learn, we can be grateful that those battles do not have to be fought, at least not on the same fields, again. We can be just as certain that others will have to be. Environmentalism or conservation and preservation, or whatever it should be called, is not a fact, and never has been. It is a job.

The idea of planning for the future is not new, but sadly, there are scant few examples of it being accomplished on even a statewide scale. Idahoans of divergent points of view still share many common values. We want to use the Snake River for agriculture and recreation. We want our public lands managed for intelligent multiple uses. We want to preserve the robust agricultural economy that the Snake River has brought into bloom. We like the solitude of open spaces, and we want to preserve the huge amounts of elbow room we enjoy on public lands.

Idahoans are more fortunate than most in that we enjoy tremendously abundant water and land resources. More than 12 million acres of public land on the Snake River Plain support a population of fewer than 850,000 people,

"Free Taters to Out of Staters" sign at the Idaho Potato Expo near Blackfoot.

Upscale subdivisions spread into the Boise foothills.

which equates to more than 16 acres of public land per capita. With an average yield of 36 million acre-feet per year, the Snake River's annual flow would flood 36 million football fields to the depth of one foot. We have abundant water to use, and statistics show that Idaho has the highest consumptive use per capita in the nation.

The issues facing the plain today will fuel political debate well into the next century. In the 1990s many decisions will be made that could determine the fate of the Snake River Plain for some time to come. How we chart the future today will largely shape our legacy. On the scale of geologic time, how will man's legacy on the plain be judged? In the long term, perhaps our 19th-century development efforts will be seen as a battle against nature, a conquest over the rugged wild landscape. In the 1900s we have witnessed a period of rapid development, of rapid growth, with a conservation ethic emerging toward the end of the century. In the next century, will we see a more harmonious relationship emerge?

Donald Worster in *Under Western Skies — Nature and History in the American West* predicts that nature, in all of its brilliant resiliency, will prevail, and so shall we. "We could never really turn all of nature into artifact. Nor could we live without nature. For all our ingenuity, we sense that we need that independent, self-organizing, resilient biophysical world to sustain us. If nature were ever truly at an end, then we would be finished. It is not, however, and we are not."

EPILOGUE
BY TODD SHALLAT

Oh faith rewarded! Now no idle dream,
The long-sought Canaan before him lies;
He floods the desert with the mountain stream,
And lo! it leaps transformed to paradise.

— *Mormon hymn*
(traditional)

DRAINING THE NORTHWESTERN Rockies and carrying more than twice the volume of the Colorado, the Snake is the nation's 10th-longest river, the

Nile of Idaho, the lifeline of the desert. Its volcanic terrain — fragile, fertile — is an odd plateau of extremes. Lewis and Clark avoided the Snake River country. The French said the river was "mad." The British discovered a sterile landscape, a barrier to civilization that might block the American claim to the disputed Northwest. So strange was the cracked plateau that Oregon-bound pioneers often dismissed southern Idaho as the barren edge of civilization, an impossible desert.

While the very strangeness of the place helped preserve its natural state, the arid steppe of the Snake, the volcanic crescent, has unleashed great expectations. What Yale historian Howard Lamar once said of the American West rings especially true on the lava

frontier: It has always been a West of the imagination, "a place to project wishes and dreams."

Thus the desert has been many landscapes. For Mormon pioneers the land was a province of Zion. For the U.S. Bureau of Reclamation the plain with its water rushing in vertical canyons was ideal for hydro projects. For modern-day conservationists the same terrain is a future dust

Centennial Marsh on the Camas Prairie: previous page, Basque shepherd; Robert W. Addison's View of Boise *(1949).*

bowl, an ecosystem under fire. Miners, ranchers, farmers, dam builders and their financiers — each wave of civilization redefined the Snake River country, each drawn by an inner compass to discover what it needed to find.

As the image of the landscape evolved, settlers held fast to a grossly inaccurate vision — the myth of southern Idaho as an empty basin, a denial of the fact that the desert had long supported a complex civilization. From the native perspective the plain was not discovered, it was invaded, and the winning of the West was much more than the "social process" that tore the nation away from Europe. The frontier was home, an actual place. Although writers often portray the Native Americans as a doomed population, the tribes have survived, and there are about as many Indians in Idaho today, about 12,000, as there were before the time of the fur trade. Clinging to a vestige of their ancestral homeland, a territory that stretched from the Tetons to the Owyhee desert in the days of the Oregon Trail, the Snake Sho-Bans retain about 1 percent of Idaho's land, including about 15 miles of riverfront property on the Fort Hall Reservation.

Indian claims to water and salmon keep frontier contests alive. In 1985, while pollution and dams seriously endangered chinook migration, the Sho-Bans, standing on treaty rights, went to federal court. "It is painful," said then-Sho-Ban attorney Larry EchoHawk; "there are so few [wild fish], our elders cannot pass on the fishing tradition." Today the wild chinook population, plunging fast, is down to a run of less than 2,000 fish.

Bitter debate over the salmon shows the new, more militant face of the Snake River environmental movement — candlelight vigils, high-stakes lawsuits, a rising intolerance for logging, farming, ranching, and the other "extractive" industries that harvest the public domain. Science is a weapon of this sometimes angry campaign, but the battles are also

fought over deeply historical issues. Critics say a century or more of ditch irrigation and grazing has been a sad history of decline. Scott Grunder, a biologist with Idaho Fish and Game, points to a 1989 state water quality study that found 4,622 miles of streams damaged and polluted by cattle grazing practices. Important fish and wildlife habitats along riparian corridors have suffered, the biologist claims. An advocate of tough new reforms — better fencing, rewards for responsible ranchers, stiff penalties for those who abuse public lands — Grunder understands that old ways die hard in the Snake River country. Still he wants pro-cattle politicians to face simple facts. "Politicians tell the ranchers that everything is okay, but it's not."

Whether or not things are mostly "okay" depends on what people believe the condition of the region was like before the Oregon Trail. "It was never a land of milk and honey," says Bert Brackett of Owyhee County, a fourth-generation rancher. Brackett tells a historical story about the range improvements that spread a thick carpet of forage across Idaho's arid plateau. Fighting science with science, the cattleman quotes from a 1989 report by the Society of Range Management, an investigation that found 79 percent of the BLM land "stable" or even "improving" in the presence of livestock grazing. "Maybe some critic's definition of good range condition is absence of domestic livestock, but," Brackett insists, "I can tell you from my experience that the range is currently in the best condition it's been in this century." Neighbors mostly agree. "Without grazing and irrigation this country would be sagebrush blackened

Barley harvest near Drummond;
following page, Swan Falls at night.

by fire," says rancher George H. Swan, a veteran of 29 years at Horse Creek Ranch south of Twin Falls.

Desert. Garden. Grassy plateau. It is easy to see how rival historical viewpoints can pose difficult questions for the future of the Snake River Plain. Will the stream vegetation recover? What about the erosion and soil depletion that scientists call desertification? Will Idaho go the way of Egypt, Iraq, arid Peru or, closer to home, the San Joaquin Valley of California — once fertile frontiers for ditch irrigation, now badly eroded, their soils poisoned by salts?

If past is prologue, as the historians say, then the relative purity of the Snake River snowmelt may spare Idaho dire effects of the toxic or briny water that plague desert farming in other parts of the globe. But Idaho agriculture has problems nevertheless. As ditch water returns to the rivers, topsoil washes into the canyons. Much of it ends up in reservoirs as silt. "Erosion from irrigation is a serious threat," says David Carter, the senior scientist at the U.S. Agricultural Research Service in Kimberly. Flying over the rich Snake River farm belt from Burley to Buhl, Carter can easily see the grey tracts of subsoil where some of North America's most fertile soil has been flushed from the fields. Carter's statistics show a direct relationship between the loss of topsoil and the loss of crops. "It takes about 15 inches of good topsoil to grow healthy crops," the scientist explains. Wheat, corn and dry beans are the crops most severely affected. Sugar beets less so. Erosion, says Carter, has hurt about 75 percent of the plain's best farmland, cutting crop production by about 25 percent.

An Idaho microchip

Scientists like Carter are quick to point out that erosion and stream degradation are not just agricultural problems. Logging, mining, road building, housing construction — even camping and boating — all bleed the land into the rivers, spoiling trout habitat and clogging the water supply. The water that flushes the soil also carries a foamy soup of chemical runoff. In 1988 a report by the U.S. Fish and Wildlife Service called American Falls Reservoir "a sump of many toxins" such as pesticides and PCBs. Downstream near Thousand Springs the Snake is a tangled swamp of rope-like filament algae. Some blame eight years of crippling drought, but river planners at the Idaho Department of Water Resources say fertilizers rich in phosphates are a major problem. Phosphates feed the aquatic plants that rob oxygen from the water and disrupt the Snake River food chain. "More rainfall alone will not fix the problem," say Ruth Schellbach, an author of the state's middle Snake management plan. Shellbach predicts that farmers in the Hagerman Valley will soon join city voters in pressing for a moratorium on river development below Milner Dam.

The farm communities that depend on these water projects take small comfort from the realization that the environmental threat to the region is an urban headache as well. Boise, for example, is booming. Its metropolitan area, 200,000 and rapidly rising, is expected to top 333,000 by the year 2010. Each year the county's three school districts need two new elementary schools just to keep pace. As the city sprawls in all directions, groundwater shrinks — not only because of more consumers, but also because houses and streets are closing the irrigation canals that seep river water into the ground. Meanwhile the rush to develop the Boise River crowds out eagles and herons. Air quality also suffers. Frightening wintertime levels of carbon monoxide, a carcinogen, have prompted wood burning bans. In 1991 an air quality group proposed a more stringent measure: "no drive" smog-alert days for older, dirtier cars.

213

Once it was easy for a sparse population to shrug off urban congestion as somebody else's concern. No longer. As the census records the inevitable shift from rural to urban, a trend in Idaho since the 1950s, farmland is cut into subdivisions and people pine for a lost way of life. You can feel the pressure of growth in a regional hub like Twin Falls, where the commercial strip crawls with five lanes of traffic and the house rental market reports a zero percent vacancy rate. You feel it in Meridian, where the city recently placed a moratorium on housing construction; in Pocatello, where the taxpayers, under legal pressure, have doubled their prison capacity with a $10.3 million jail; in Victor near the Wyoming border, where half-acre lots that recently sold for $2,000 now quickly sell for 10 times that amount; in Idaho Falls, where the Salvation Army reports a sharp rise in the homeless and a seven-bed Baptist mission is the only shelter for single men. And you can feel the pinch of the future in quiet places like Kuna with its four new subdivisions. "We're not Boise yet," says Ann Danes of Kuna, a city clerk who reports "mixed feelings" toward growth. "You still don't have to look at the sidewalk when you pass people, and you can still come into city hall to talk about grandchildren and quilt-making." With regret in her voice she adds: "We're going to lose all that."

While urban sprawl may be the future of the Snake River country, there are still free-flowing rivers and open range that, when compared to the rest of the nation, give Idahoans great reason for hope. Idaho's largest cities are still remarkably clean. There are still brilliant skies and magnificent vistas. And on the eastern edge of the plain, where the South Fork of the Snake moves west from the Yellowstone country, a hiker can still scale the cracks of Dry Canyon where the plain once oozed with basalt.

Shade your eyes against the glare of the sun and take in the blank expanse of the mysterious landscape. Squinting toward Idaho Falls, you can almost picture the plain as the Shoshone once saw it — pine and cottonwood streams cutting through amber grasses, a sea of shimmering prairie swept by ecological change.

Bob Limbert publicity photo, about 1930; following pages (from the BSU Limbert Collection): Limbert postcard of hunter with grizzly, about 1915; canyon of the Snake; Melba-area homesteaders in lava rock cabin, about 1930; rifleman at Craters of the Moon; "Valley of the Moon" panorama, about 1921; porcupine; beaver; owl; Canyon of Brown in Owyhee County, 1922; Limbert's taxidermy shop with crated elk head, a gift to President Warren G. Harding, 1921; stuffing a rattlesnake; Limbert on motorcycle at Map Rock near Marsing, about 1922; Limbert's display of Idaho products at the San Francisco World's Fair, 1915; wrestling over the Snake.

"TWO GUN" BOB: IDAHO'S SHOWMAN

Publicists called him the rightful heir to Bat Masterson and Wyatt Earp, a quick-draw gunman who from a distance of 25 feet could split a bullet on the edge of an ax, or toss a bottle into the air and shoot through its cork and bottom without breaking the sides. To Al Capone he was the cowboy from Boise who captured Chicago headlines by calling the gangsters "starched-up little pets" and challenging them all to a gunfight. To *Sunset Magazine*, reporting on Idaho in 1924, he was the adventurer who filmed and photographed a Snake River wonderland of lava flows, ice caves, natural bridges and Indian petroglyph fields. To Calvin Coolidge and the U.S. Geological Survey the same showman-naturalist was a serious man of science and the driving force behind the presidential proclamation that created a national monument at Craters of the Moon.

Robert W. Limbert saw Idaho as a fantasyland and packaged it for the nation. A visionary, he used photography and popular science to picture the state as a tourist mecca in an era of rising nostalgia for America's lost frontier.

Born in 1885 and raised in Omaha, Nebraska, Limbert moved to Boise at age 26 to enter the taxidermy business. In 1915 his bizarre collection of stuffed animals and other Idaho curios —

ACKNOWLEDGEMENTS

Like the majestic river, our study has evolved slowly with time. What began in 1987 as a scholarly anthology has grown into a more ambitious attempt to penetrate and illustrate the human encounter with nature. Over its long gestation we were sustained by a shared fascination for the mystery of the Snake River country and also by the conviction that a wide-ranging book of this kind was consistent with Boise State's long-standing commitment to Western studies.

Too many people have contributed to this work to thank them all individually here, but some deserve special mention. Editors Todd Shallat, Larry Burke, Chris Latter and Josie Fretwell brainstormed the graphics, wrote capsules and captions, edited and reworked the copy, compiled the sources and index, and designed and produced the book. Dave Clark first conceived of the project and then served as our science consultant and liaison to the Craters of the Moon Natural History Association. Robert Sims, former dean of the BSU College of Social Sciences and Public Affairs, worked with Shallat and Clark to organize the project and broaden its scope through applied social science. Former BSU President John Keiser, former research director Jim Baker, and former interim President Larry Selland were quick to recognize that the largest university on the Snake River Plain had a special obligation to study and interpret the region. We also appreciate generous support from President Charles Ruch.

No amount of encouragement would have seen this project through to completion without an enormous investment of time from the hard-working staff at BSU News Services. Glenn Oakley co-authored a chapter, helped with the geology and geography captions, and photographed much of the book. Chuck Scheer drew more than a dozen maps and photographed the artifacts on loan from museums. Amy Stahl, Bob Evancho and Brenda Haight worked on the copy. Arianne Poindexter entered corrections and changes on PageMaker. Special thanks also go to our gifted artist-illustrator Gwenetta Bates, photographer Steve Bly and copy editors Bob Fullilove and Sandy Marostica.

In our search for information and graphics we sought Idaho experts who were generous with their knowledge and time. Merle Wells, Jim Woods, Peter Bowler, Max Pavesic, Margaret Jenks and Hugh Lovin made essential contributions. And no book of this kind could have been produced without help from the friendly staff of the Idaho State Historical Society. We also appreciate the support of the Idaho Museum of Natural History at Idaho State University and the Herrett Museum at College of Southern Idaho. We are especially indebted to Guila Ford, Ken Swanson, Bill Tydeman, Thomas Jaehn, Jody Hawley Ochoa, Joe Toluse, John Yandell, Elizabeth Jacox, Larry Jones and Gary Bettis. At the Boise State Library Special Collections we relied on Mary Carter and Alan Virta. David Kennedy was our research assistant during the first year of the project. Vital assistance was also provided by Karen Kelsch, Dan Meatte, Arthur Hart, Kathleen Bettis, Chuck Jones, Bob Sutter, Jim Peaco, Tom Ansbach, Tom Trusky, Jean Wilson, Denise Nelis, Marilyn Paterson, Barbara Valdez, John and Donna Isaacs, Ann Van Buren, Phillip McClain, Jerry Lindholm, Trina Olson, Diane Ronayne, Jim Peaco, Wendy Downs, Sharon Holmer, Madeline Buckendorf, Cort Conley, Tom Green, Mary Suter, Dana Quinney, Alan Robertson, John Killmaster, Susan Stacy, Neil King and Pat Costello.

We also gratefully acknowledge support from our friends in the Boise State History Department who read portions of the manuscript and led us to sources and concepts that greatly enriched our book.

Finally, we appreciate the high quality work performed by the staff at both Hi-Tech Color and Lithocraft, Inc. in the production of the book.

including an 11-foot model of a Russet Burbank potato — won acclaim and a sack full of medals at San Francisco's international expo.

Collector, fiction writer, filmmaker and dude-ranch entrepreneur, Limbert was known most of all for astonishing work with a Graflex camera. Capturing a region still largely untouched by tourism and urbanization, his photos of wildlife, Indians, rock art, mountains and vertical canyons — more than 8,000 pictures in all — are a rich archive of information that help modern scholars assess environmental changes on the plain. Today that photographic record, together with many of Limbert's pamphlets and papers, is open to researchers at the Boise State University Library Special Collections Department.

SOURCES/CREDITS

SELECTIVE BIBLIOGRAPHY AND SOURCES OF THE QUOTATIONS

CH. 1: SEEING THE LAND

Alt, David D., and Donald W. Hyndman, *Roadside Geology of Idaho* (1989); **Arrington**, Leonard J., "Irrigation In the Snake River Valley: An Historical Overview," *Idaho Yesterdays* 30 (Summer/Spring 1986):3-11 ["They thought I was crazy" quote from p. 8]; **Bartlett**, Richard, *Great Surveys of the American West* (1962); **Buwulda**, John P., "A Preliminary Reconnaissance of the Gas and Oil Possibilities of Southwestern and South-Central Idaho," *Idaho Bureau of Mines and Geology*, pamphlet 5 (July 1923) ["unfavorable" quote from p. 10]; **Gazin**, C. Lewis, "Fossil Hunting in Southern Idaho," *Smithsonian Institute Exploration and Field Work*, no. 3300 (1934); **Gertsch,** Darrell, "Contours of Change: Water Resource Allocation and Economic Stability in the Snake River Basin," *Idaho Yesterdays* 30 (Summer/Spring 1986):12-19; **Hackett**, Bill, Jack Pelton, and Chuck Brockway, *Geohydrologic Story of the Eastern Snake River Plain and the Idaho National Engineering Laboratory* (November 1986); **Irving,** Washington, *The Adventures of Captain Bonneville, U.S.A., in the Rocky Mountains and the Far West* (1849) ["forlorn … fantastic," "volcanic character" quotes from pp. 254-57, 261]; **King**, Clarence, *Mountaineering in the Sierra Nevada* (1903) ["blackened ruins" quote from p. 243]; **Kirkham**, Virgil R.D., "Oil Possibilities and Drilling Activities in South Idaho," *Idaho Engineer* 2:1 (1925):11-14; **Maley**, Terry, *Exploring Idaho Geology* (1987); **Plew**, Mark G., *An Introduction to the Archaeology of Southern Idaho* (1986); **Raymond**, Rossiter W., *Mineral Resources of the States and Territories West of the Rocky Mountains* (1869), excerpted in "The Report of the Special Commissioner," *Idaho Yesterdays* 33 (Fall 1989) ["handwriting of nature" quote from p. 26]; **Reed**, Scott W., "The Other Uses of Water," *Idaho Yesterdays* 30 (Summer/Spring 1986) ["unholy alliance" quote from p. 38]; **Robinson**, Michael C., *Water for the West: The Bureau of Reclamation, 1902-1977* (1979); **Russell**, Israel C., "Geology and Water Resources of the Snake River Plains

in Idaho," *U.S. Geological Survey Bulletin,* no. 199 (1902); **State of Idaho,** Dept. of Agriculture, *1989 Agriculture Statistics* (1989); **Stearns**, Harold T., "Origins of the

Large Springs and Their Alcoves Along the Snake River in Southern Idaho," *Journal of Geology* 44 (1936):429-50; Stearns, "Snake River—Idaho's Greatest Resource," *Idaho Power Company Bulletin* 67 (May 1956):3-5; **Swanson**, Earl H., Jr., "The Snake River Plain," *Idaho Yesterdays* 18 (Summer 1974):2-11.

CH. 2: VOLCANIC CRESCENT

Christiansen, R.L., "Yellowstone Magmatic Evolution: Its Bearing on Understanding Large-volume Explosive Volcanism," *Explosive Volcanism: Inception, Evolution, Hazards* (1984):84-95; **Christiansen**, R.L., and H.R. Blank, Jr., "Volcanic Stratigraphy of the Quaternary Rhyolite Plateau in Yellowstone National

Park," *U.S. Geological Survey Professional Paper*, no. 729-B (1972); **Creighton**, D.N., "Menan Buttes, Southeastern Idaho," Rocky Mountain Section of the Geological Society of America, *Centennial Field Guide*, 2 (1987):109-11; **Doherty**, D.J., L.A. McBroome, and M.A. Kuntz, "Preliminary Geologic Interpretation and Lithologic Log of the Exploratory Test Well (INEL-1), Idaho National Engineering Laboratory, Eastern Snake River Plain, Idaho," *U.S. Geological Survey Open File Report*, no. 79-1248 (1979); **Hackett**, W.R., and L.A. Morgan, "Explosive Basaltic and Rhyolitic Volcanism of the Eastern Snake River Plain," *Guidebook to the Geology of Central and Southern Idaho: Idaho Geological Survey Bulletin*, P.K. Link and W.R. Hackett, eds., no. 27 (1988):283-301; **Howard**, K.A., J.W. Shervais, and E.H. McKee, "Canyon-filling Lavas and Lava Dams on the Boise River, Idaho, and Their Significance for Evaluation Downcutting During the Last Two Million Years," *Cenozoic Geology of Idaho: Idaho Bureau of Mines and Geology Bulletin*, Bill Bonnichsen and R. M. Breckenridge, eds., no. 26 (1982):629-41; **King**, J.S., "Selected Volcanic Features of the South-Central Snake River Plain, Idaho," *Cenozoic Geology of Idaho: Idaho Bureau of Mines and Geology Bulletin*, no. 26 (1982):439-52; **Kuntz**, M.A., "Geology of the Arco-Big Southern Butte Area, Eastern Snake River Plain, and Potential Volcanic Hazards to the Radioactive Waste Management Complex and Other Waste Storage and Reactor Facilities at the Idaho National Engineering Laboratory, Idaho," *U.S. Geological Survey Open File Report*, no. 78-691 (1978); Kuntz, "Geologic Map of the Arco-Big Southern Butte area, Butte, Blaine and Bingham Counties, Idaho," (1:48,000 scale) *U.S. Geological Survey Open-File Report*, no. 78-302 (1978); **Kuntz**, M.A., and G.B. Dalrymple, "Geology, Geochronology and Potential Volcanic Hazards in the Lava Ridge-Hell's Half Acre Area, Eastern Snake River Plain, Idaho," *U.S. Geological Survey Open-File Report*, no. 79-1657 (1979); **Kuntz** et al., "Geologic Map of the Lava Ridge-Hell's Half Acre Area, Eastern Snake River Plain, Idaho," (1:48,000 scale) *U.S. Geological Survey Open-File Report*, no.

79-1657 (1979); **Kuntz** et al., "Contrasting Magma Types and Steady-state, Volume-predictable Volcanism Along the Great Rift, Idaho," *Geological Society of America Bulletin*, 97 (1986):579-94; **Kuntz** et al., "Geologic Map of the Craters of the Moon, Kings Bowl, and Wapi Lava Fields, and the Great Rift Volcanic Rift Zone, South-central Idaho," (1:1000,000 scale) *U.S. Geological Survey Miscellaneous Investigations Series*, map I-1632 (1988); **Malde**, H.E.,

"The Catastrophic Late Pleistocene Bonneville Flood in the Snake River Plain," *U.S. Geological Survey Professional Paper*, no. 596 (1968); Malde, "The Yahoo Clay, a Lacustrine Unit Impounded by the McKinney Basalt in the Snake River Canyon Near Bliss, Idaho," *Cenozoic geology of Idaho: Idaho Bureau of Mines and Geology Bulletin*, no. 26 (1982):617-28; **Morgan**, L.A., "Explosive Rhyolitic Volcanism on the Eastern Snake River Plain, Idaho," (Ph.D. diss., University of Hawaii, Manoa, 1988); **Morgan**, L.A., D.J. Doherty, and W.P. Leeman, "Ignimbrites of the Eastern Snake River Plain: Evidence for Major Caldera-forming Eruptions," *Journal of Geophysical Research* 89 (1984):8665-78; **Whitehead**, R.L., "Geohydrologic Framework of the Snake River Plain, Idaho and Eastern Oregon," (1:1,000,000 scale) *U.S. Geological Survey Hydrologic Investigations Atlas*, no. HA-681 (1986); **Womer**, M.B., Ronald Greeley, and J.S. King, "Phreatic Eruptions of the Eastern Snake River Plain of Idaho," *Cenozoic Geology of Idaho: Idaho Bureau of Mines and Geology Bulletin*, no. 26 (1982):453-64; **Wood**, W.W., and W.H. Low, "Aqueous Geochemistry and Diagenesis in the Eastern Snake River Plain Aquifer System, Idaho," *Geological Society of America Bulletin* 97 (1986):1456-66.

CH. 3: A CLIMATE OF CHANGE

Atmospheric Environment, Environment Canada, *Canadian Normals, Temperature* (1973); **Baker**, Charles, interviewed by Glenn Oakley (April 4, 1993); **Barker**, R.J., Robert E. McDole, and Glen H. Logan, *Idaho Soils Atlas* (1983); **Bechard**, Marc, interviewed by Glenn Oakley (March 15, 1993); **Bentley**, E.B., "Glacial

Morphology of Eastern Oregon Uplands" (Ph.D. diss., University of Oregon 1974); **Boone**, Lalia, *Idaho Place Names: A Geographical Dictionary* (1988); **Bowler**, Peter A., "Natural History Studies and an Evaluation for Eligibility of the Wiley Reach of the Snake River for National Natural Landmark Designation" (typewritten, 1981); **Bowler**, Peter A., "Natural History Studies and an Evaluation for Eligibility of Malad Canyon for National Natural Landmark Designation" (typewritten, 1981); **Bright**, R.C., "Pollen and Seed Stratigraphy of Swan Lake, Southern Idaho: Its Relation to Regional Vegetational History and Lake Bonneville History," *Tebiwa* 9:2 (1966):1-47; **Broeker**, W.S., and P.C. Orr, "Radiocarbon Chronology of Lake Lahontan and Lake Bonneville," *Geological Society of America Bulletin* (1958):1009-

32; **Burt**, W.H., and R.P. Grossenheider, *A Field Guide to the Mammals*, Peterson Field Guide Series (1978); **Bureau of Land Management**, Idaho Falls District, and U.S. Forest Service, Targee National Forest, *Snake River Activity-Operations Plan Environmental Assessment* (February 1991) ["number of young trees being established" quote from p. 44]; **Butler**, B. Robert, *Idaho Archaeology: The Upper Snake and Salmon River Country* (1978); **Conley**, Cort, *Idaho for the Curious: A Guide* (1982); **Connelly**, Jack, "Sage Grouse and Fire: A Complex Issue for Resource Management," *Idaho Wildlife* (September/Octo-

ber 1988); **Cunningham**, G.D., "Hagerman Fauna Area Paleontological Survey," no. ID-010-CT3-17, *U.S. Bureau of Land Management*, Boise District (1984):1-56; **Davidson**, Abraham A., "The Wretched Life and Death of an American Van Gogh," *Smithsonian* (December 1987); **Devoto**, Bernard, *Across the Wide Missouri* (1975); **Frest**, T.J., and P.A. Bowler, "A Preliminary Checklist of the Aquatic and Terrestrial Mollusks of the Middle Snake River Sub-Basin," *Proceedings of the Desert Fishes Council* (1993); **Hagerman Fossil Beds National Monument,** "Evaluation of Hagerman Fauna Sites, Twin Falls, Idaho for Eligibility for Registered Natural Landmark Designation and Natural Landmark Brief," (March 26, 1975) by R.W. Jones; **Henry**, Craig, "Holocene Paleoecology of the Western Snake River Plain, Idaho," (Ph.D. diss., University of Michigan, 1984); **Heusser**, C.J., "Late-Pleistocene Environments of North Pacific North America," *American Geographical Society*, Special Publication no. 35 (1960); **Howard**, Rich, interviewed by Glenn Oakley (February 2, 1993); **Hunt**, Charles B., *Natural Regions of the United States and Canada* (1974); **Huntley**, James L., *Ferryboats in Idaho* (1979); **Irving**, Washington, *The Adventures of Captain Bonneville*, ed. Robert A. Rees and Alan Sandy (1977); **Jarrett**, R.D., and H.E. Malde, "Paleodischarge of the Late Pleistocene Bonneville Flood, Snake River, Idaho, Computed from New Evidence," *Geological Society of America Bulletin* 99 (1987):127-34; **Kjelstrom**, L.C., and R.L. Moffat, "Method of Estimating Flood-Frequency Parameters for Streams in Idaho," *U.S. Geological Survey Open File Report*, no. 81-0909 (1982); **MacMahon**, J.A., *Deserts*, Audubon Society Nature Guides (1985); **Malde**, H.E., "The Catastrophic Late Pleistocene Bonneville Flood in the Snake River Plain, Idaho," *U.S. Geological Survey Professional Paper*, no. 596 (1968); **Merigliano**, Mike, interviewed by Glenn Oakley (February 10, 1993); **Miller**, Susanne J., "The Archaeology and Geology of an Extinct Megafauna/Fluted-Point Association of Owl Cave, the Wasden Site, Idaho: A Preliminary Report," *Peopling of the New World*, Jonathon E. Erickson et al., eds., Ballena Press Anthropological Papers, no. 23 (1982):81-95; **Naderman**, Justin, interviewed by Glenn Oakley (February 11, 1992, and March 20, 1993; **Palmer**, Tim, *The Snake River: Window to the West* (1991); **Plew**, Mark C., *An Introduction to the Archaeology of Southern Idaho* (1986); **Rue**, Leonard Lee, *Game Animals* (1968); **Rhodenbaugh**, Edward F., *Sketches of Idaho Geology* (1961); **Ronayne**, Diane, "Enchanted Circle," *Northern Lights* (November/December 1987): 30-4; **Rosentreter**, Roger, interviewed by Glenn Oakley (March 15, 1993); **Russell**, Israel C., "A Geological Reconnaissance in Southern Oregon," *U.S. Geological Survey 4th Annual Report* (1882):435-64; Russell, "Preliminary Report on Artesian Basins in Southwestern Idaho and Southeastern Oregon," *U.S. Geological Survey Water Supply Paper*, no. 78 (1903); Russell, "Notes on the Geology of Southwestern Idaho and Southeastern Oregon," *U.S. Geological Survey Bulletin*, no. 217 (1903); **Russell**, Osborne, *Journal of a Trapper* (1955); **Saab**, Vicky, interviewed by Glenn Oakley (February 20, 1993); **Scott**, W. Frank, *Potential Natural Landmarks, Geologic Themes, on the Columbia Plateau* (1978); **Scott**, William E., et al., "Reinterpretation of the Exposed Record of the Last Two Cycles of Lake Bonneville, Western United States," *Quaternary Research* 20:3 (1983):261-85; Scott, et al., "Revised Quaternary Stratigraphy and Chronology in the American Falls Area, Southeastern Idaho," *Cenozoic Geology of Idaho: Idaho Bureau of Mines and Geology Bulletin*, no. 26 (1982):581-95; **Smith**, G.R., "Fishes of the Pliocene Glenns Ferry Formation, Southwest Idaho," *Papers on Paleontology No. 14, Claude W.*

Hibbard Memorial Volume 5 (1975); **State of Idaho**, Fish and Game Dept., *Species Management Plan* (1985); **Tourism and Industrial Development**, Idaho Division, *Idaho Almanac* (1977); **Trimble**, Stephen, *The Sagebrush Ocean* (1989); **Udvardy**, M.D. F., and S. Rayfield, *The Audubon Society Field Guide to North American Birds, Western Region* (1977); **U.S. Bureau of Land Management**, Jarbidge Resource Area, "Natural History Resource Management Plan, Hagerman Fauna Sites National Natural Landmark," (1968) by T.R. Weasma; **U.S. Interior Dept.**, Bureau of Land Management, *Borah-Midpoint 500 KV Transmission Line Environmental Assessment Report* (1979); **U.S. Interior Dept.**, Bureau of Reclamanation. "Preliminary Geologic Report on the Bliss Dam Site, Bruneau Project, Idaho," (1950), by L.D. Jarrard.

CH. 4: NATIVE TRADITIONS

Agenbroad, Larry D., "Buffalo Jump Complexes in Owyhee County, Idaho," *Miscellaneous Papers of the Idaho State University Museum of Natural History*, no. 1 (December 1976):1-38; **Ames**, Kenneth M., "Diversity and Variability in the Prehistory of Southwestern Idaho," *Idaho Archaeologist* 6 (1982):1-10; **Bonnichsen**, Robson, "The Rattlesnake Canyon Cremation Site, Southwestern Idaho," *Tebiwa* 7:1 (1964):28-38; **Bright**, Robert C., "Pollen and Seed Stratigraphy of Swan Lake, Southeastern Idaho: Its Relation to Regional Vegetational History and to Lake

Bonneville History," *Tebiwa* 9:2 (1966):1-47; **Butler**, B. Robert, "A Bison Jump in the Upper Salmon River Valley," *Tebiwa* 14:1 (1971):4-32; Butler, *A Guide to Understanding Idaho Archaeology: The Upper Snake and Salmon River Country* (1978); Butler, "The Native Pottery of the Upper Snake and Salmon River Country," *Idaho Archaeologist* 3:1 (1979):1-10; Butler, "Prehistory of the Snake and Salmon River Area," *Handbook of North American Indians: Great Basin*, Warren L. D'Azevedo, ed., vol. 11 (1986):127-34; **Butler**, B. Robert, Helen Gildersleeve, and John Sommers, "The Wasden Site Bison: Sources of Morphological Variation," *Aboriginal Man and Environments on the Plateau of Northwest America*, A.H. Stryd and R.A. Smith, eds. (1971):126-52; **Cinadr**, Thomas J., "Mount Bennett Hills Planning Unit: Analysis of Archaeological Resources," *Archaeological Reports*, no. 6 (1976); **Crabtree**, Don E., "Archaeological Evidence of Acculturation Along the Oregon Trail," *Tebiwa* 11:2 (1968):38-42; **Davis**, John C., Paul R. Fenske, and H. Thomas Ore, "Geological Evaluation of the Haskett Site," *Tebiwa* 8:2 (1965):29-32; **Farb**, Peter, "The Birth and Death of the Plains Indians," *Man's Rise to Civilization as Shown by the Indians of North America from Primeval Times to the Coming of the Industrial State* (1968); **Green**, James P., "Archaeology of the Rock Creek Site, 10-CA-33, Sawtooth National Forest, Cassia County, Idaho," (Master's thesis, Idaho State University, 1972); **Green**, Thomas J., "House Form and Variability at Givens Hot Springs, Southwest Idaho," *Idaho Archaeologist* 6:1-2 (1982):33-44; **Gruhn**, Ruth,

"The Archaeology of Wilson Butte Cave, South-Central Idaho," *Occasional Papers of the Idaho State College Museum*, no. 6 (1961); Gruhn, "A Collection of Artifacts from Pence-Duerig Cave in South-Central Idaho," *Tebiwa* 4:1 (1961):1-24; Gruhn, "Test Excavations at Sites 10-OE-128 and 10-OE-129, Southwest Idaho," *Tebiwa* 7:2 (1964):28-36; **Jennings**, Jesse D., "Danger Cave," *Memoirs of the Society For American Archaeology*, no. 14 (1957); **Lahren**, Larry, and Robson Bonnichsen, "Bone Foreshafts from a Clovis Burial in Southwestern Montana," *Science* 186 (1974):147-50; **Lawrence**, Barbara, "Antiquity of Large Dogs in North America," *Tebiwa* 11:2 (1968):43-9; **Liljeblad**, Sven, "Indian Peoples of Idaho," (Manuscript on file, Idaho State University Museum, 1957); **Lynch**, Thomas F., and Lawrence Olsen, "The Columbet Creek Rockshelter," *Tebiwa* 7:1 (1964):7-16; **Meatte**, Daniel S., "Two Archaic Fish Traps from the Snake River Canyon near Twin Falls, Idaho," *Idaho Archaeologist* 9:2 (1986):15-9; **Metzler**, Sharon, "The Brown Creek Archaeological Survey, Owyhee County, Idaho," *Archaeological Reports*, no. 2 (1976); Metzler, "An Archaeological Survey of Castle Creek, Idaho," (Report on file, Boise District Office, Bureau of Land Management, 1977); **Miller**, Susanne J., "The Archaeology and Geology of an Extinct Megafauna/Fluted Point Association at Owl Cave, the Wasden Site, Idaho: A Preliminary Report," *Peopling of the New World*, Jonathan E. Ericson et al., eds., Ballena Press Anthropological Papers, no. 23 (1982):81-95; **Miller**, Susanne J., and Wakefield Dort, Jr., "Early Man at Owl Cave: Current Investigation at the Wasden Site, Eastern Snake River Plain, Idaho," *Early Man in America from a Circum-Pacific Perspective*, Allan Lyle Bryan, ed. (1978); **Murphey**, Kelly, "An Archaeological Inventory of Devils Creek, Owyhee and Twin Falls Counties, Idaho," *University of Idaho Anthropological Research Manuscript Series*, no. 35 (1977); **Murphey**, Robert F. and Yolanda Murphey, "Shoshone-Bannock Subsistence and Society," *Anthropological Records* 16 (1960):293-338; **Pavesic**, Max G., "Cache Blades and Turkey Tails: Piecing Together the Western Idaho Archaic Burial Complex," *Stone Tool Analysis: Essays in Honor of Don E. Crabtree*, Mark G. Plew et al., eds. (1985):55-89; Pavesic, "Archaeological Evidence for Anadromous Fish Use in Bannock-Shoshoni Territory," (Affidavit Prepared for the Sho-Ban Tribes of Idaho, Inc., *U.S. vs. Oregon* Intervention Hearing, U.S. District Court, Portland, 1986); **Pavesic**, Max G., and Daniel S. Meatte, "Archaeological Test Excavations at the National Fish Hatchery Locality, Hagerman Valley, Idaho," *Archaeological Reports*, no. 8 (1980); **Pavesic**, Max G., W.I. Follett, and William P. Statham, "Anadromous Fish Remains from Schellbach Cave No. 1, Southwestern Idaho," *Idaho Archaeologist* 10:2 (1987):23-6; **Penson**, Betty, "Idaho Archaeologist Carves Out National Reputation," *Idaho Statesman* (May 13, 20, 1979); **Petersen**, Nicholas H., "A Clovis Point from Long Valley, Idaho," *Idaho Archaeologist* 10:2 (1987):41-2; **Plew**, Mark G., "An Archaeological Inventory Survey of the Camas Creek Drainage Basin, Owyhee County, Idaho," *Archaeological Reports*, no. 1 (1976); Plew, "Aboriginal Hunting Complexes in the Owyhee Uplands, Idaho," *The Masterkey* 53 (1979):108-11; Plew, "Southern Idaho Plain: Implications for Fremont-Shoshoni Relationships in Southwestern Idaho," *Plains Anthropologist* 24 (1979):329-35; Plew, "Fish Remains from Nahas Cave: Archaeological Evidence of Anadromous Fishes in Southwestern Idaho," *Journal of California and Great Basin Anthropology* 2 (1980):129-32; Plew, "Archaeological Investigations in the Owyhee Uplands, Idaho," *Archaeological Reports*, no. 7 (1980); Plew, "Archaeological Excavations at Big Foot Bar, Snake River Birds of Prey Natural Area, Idaho," *Project Reports*, no. 3, Idaho Archaeological Consultants (1980); Plew, "Archaeological Test Excavations at Four Prehistoric Sites in the Western Snake River Canyon near Bliss, Idaho," *Project Reports*, no. 5, Idaho Archaeological Consultants (1981); Plew, "A Preliminary Overview of the Owyhee Country," *Idaho Archaeologist* 6:1-2 (1982):47-54; Plew, "Implications of Nutritional Potentials of Anadromous Fish Resources of the Western Snake River Plain," *Journal of California and Great Basin Anthropology* 5:1-2 (1983):58-65; Plew, *An Introduction to the Archaeology of Southern Idaho* (1986); Plew, "The Archaeology of Nahas Cave: Material Remains and Chronology," *Archaeological Reports*, no. 13 (1986); Plew, "A Reassessment of the Five Finger and "Y" Buffalo Jumps, Southwest Idaho," *Plains Anthropologist* 32:117 (1987):317-21; Plew, "Archaeological Assemblage Variability in Fishing Locales of

the Western Snake River Plain," *North American Archaeologist* 9:3 (1988):247-57; Plew, "Archaeological Investigations along the East and South Forks of the Owyhee River," *Idaho Archaeologist* 6:1-2 (1982):25-32; Plew, "Test Excavations at the Kueney Site (10 TF 327): A Middle Archaic Site in the South Hills Country," *Idaho Archaeologist* 8:2 (1985): Plew, "Archaeological Explorations in Southwestern Idaho," *American Antiquity* 31:1 (1965):24-37; Plew, "Archaeological Exploration of The Snake River Plain," *Idaho Yesterdays* 18:2(1974):13-14; Plew, *Birch Creek: Human Ecology in the Cool Desert of the Northern Rock Mountains, 9000 B.C.–A.D. 1850* (1972); **Plew**, Mark G., Kenneth M. Ames, and Christen K. Fuhrman, "Archaeological Excavations at Silver Bridge (10-BO-1), Southwest Idaho," *Archaeological Reports*, no. 12 (1984); **Plew**, Mark G., James C. Woods, and Max G. Pavesic, *Stone Tool Analysis: Essays in Honor of Don Crabtree* (1985); **Plew**, Mark G., and Kevin Meyer, "An Aboriginally Worked Brass Bipoint from Three Island Crossing," *Idaho Archaeologist* 10:1 (1987):17-18; **Plew**, Mark G., Max G. Pavesic, and Mary Anne Davis, "Archaeological Investigations at Baker Caves I and III: A Late Archaic Component on the Eastern Snake River Plain," *Archaeological Reports*, no. 15 (1987); **Sappington**, Robert Lee, "The Archaeology of the Lydle Gulch Site (10-AA-72) Prehistoric Occupation in the Boise River Canyon, Southwestern Idaho," *University of Idaho Research Manuscript Series*, no. 66 (1981); **Sargeant**, Kathryn E., "The Haskett Tradition: A View from Redfish Overhang," (Master's thesis, Idaho State University, 1973); **Schellbach**, Louis, "The Excavation of Cave No. 1, Southeastern Idaho," *Tebiwa* 10:2 (1967):63-72; **Steward**, Julian H., "Basin-Plateau Socio-Political Groups," *Bureau of American Ethnology Bulletin*, no. 120 (1938); **Swanson**, Earl H., Jr., "Folsom Man in Idaho," *Idaho Yesterdays* 5:1 (1961):26-30; **Swanson**, Earl H., Jr., and Paul G. Sneed, "Jacknife Cave," *Tebiwa* 14:1 (1971):33-69; **Titmus**, Gene L., "The Blue Lakes Clovis," *Idaho Archaeologist* 10:2 (1987):39-40; **Woods**, James C., and Gene L. Titmus, "A Review of the Simon Clovis Collection," *Idaho Archaeologist* 8:1 (1985):3-8.

CH. 5: CONFRONTING THE DESERT

Alegria, Henry, *75 Years of Memoirs* (1981); **Athearn**, Robert G., "The Oregon Short Line," *Idaho Yesterdays* 13 (Winter 1969-70):2-18; Athearn, *Union Pacific Country* (1971); **Attebery**, Louie W., ed., *Idaho Folklife: Homesteads to Headstones* (1985); **Beal**, Merrill D., "The Story of the Utah Northern Railroad," *Idaho Yesterdays* 1 (Spring 1957):3-20, (Summer 1957):16-23; **Beal**, Merrill D., and Merle W. Wells, *History of Idaho*, 3 vols., (1959); **Beck**, Richard J., *Famous Idahoans* (1989); **Bek**, William G., trans., "From Bethel, Missouri, to Aurora, Oregon: Letters of William Keil, 1855-70," *Missouri Historical Review* 47 (October 1953) ["seventh prince ... hideous world" quote from pp. 33-34]; **Boise**: The Illustrated Idaho Co., *Boise Idaho: Forty-five Years of Progress* (1911) ["imperial city" quote from p. 1]; **Brink**, Helen C.G., "Toms and Hens," *Here's Idaho!* (January 1933); **Butler**, B. Robert, *When Did the Shoshoni Begin to Occupy Southern Idaho?* (1981); **Buckendorf**, Madeline, interviewed by Todd Shallat (July 29 and 30, 1991); **Byington**, Fay Kofoed, *History of Lava Hot Springs, Idaho Our Valley* (1989); **Churchill**, Claire Warner, ed., "The Journey to Oregon— A Pioneer Girl's Dairy," *Oregon Historical Quarterly* 29 (March 1928) ["oxen died ... He has lost two" quote from p. 92]; **Corless**, Hank, *The Weiser Indians: Shoshoni Peacemakers* (1990); **Davidson**, James W., et al., *Nation of Nations: A Narrative History of the American Public* (1990) ["endure heat like a Salamander" quote from p. 497]; **Drury**, Clifford Merrill, *First White Woman Over the Rockies: Diaries, Letters, and Biographical Sketches of the Six Women of the Oregon Mission Who Made the Overland Journey in 1836 and 1838*, 3 vols.

(1963-66) ["Heavens over us ... many mercies" quote from p. 218]; **Frémont**, John C., *Report of the Exploring Expedition to the Rocky Mountains* (1966) ["Beyond this place," "nutritious" quotes from pp. 261, 273]; **Foote**, Mary Hallock, *A Victorian Gentlewoman in the Far West*, Rodman Paul, ed. (1972) ["courtesy call," "darkest Idaho ... empty of history" quotes from pp. 26, 265]; **Fuller**, Margaret, *Trails of Western Idaho from Sun Valley to Hells Canyon* (1982); **Gertsch**, W. Darrell, "The Upper Snake River Project: A Historical Study of Reclamation and Regional Development, 1890-1930," (Ph.D. diss., University of Washington, 1974); Gertsch, "Water Use, Energy, and Economic Development in the Snake River Basin," *Idaho Yesterdays* 23 (Summer 1979):58-72; **Hart**, Arthur A., *The Boiseans: At Home* (1984); Hart, *Camera Eye on Idaho: Pioneer Photography* (1990); **Hayden**, Ferdinand V., *Sixth Annual Reports of the United States Geological Survey of the Territories, Embracing Portions of Montana, Idaho, Wyoming, and Utah: Being a Report of Progress of the Explorations for the Year, 1872* (1872) ["There seems to be no want of fertility" quote p. 203]; **Idaho Bureau of Mines and Geology**, *The Geology and Mineral Resources Section in Mineral and Water Resources of Idaho* (1982); **Idaho Centennial Commission**, *Idaho's Ethnic Heritage: Historical Overviews* (1990); **Idaho Statesman**, "Census Decision Stirs Idaho Hispanics," (July 16, 1991); **Irving**, Washington, *Astoria; or, Anecdotes of an Enterprise Beyond the Rocky Mountains* (1868) ["burnt and barren," "barren rocky," "wide sunburnt," "dismal desert," "cheerless wastes" quotes from pp. 389, 518, 403, 397, and 529]; Irving, *The Adventures of Captain Bonneville, U.S.A., in the Rocky Mountains and the Far West* (1961) ["sandy and volcanic ... irreclaimable" quote from p. 212]; **Jackson**, Donald, and Mary Lee Spence, eds., *The Expeditions of John Charles Frémont*, 3 vols. (1970-1980) ["dark ... contrasted effect" quote from p. 535]; **Jeffrey**, Julie Roy, *Frontier Women: The Trans-Missisippi West, 1840-1880* (1979); **Jerome Commercial Club**, *Jerome and Its Environment* (n.d., mss 544, box 5, pam 12, Idaho State Archives) ["No land can boast" quote from p. 3]; **Jones**, Dorsey D., ed., "Two Letters by a Pioneer from Arkansas," *Oregon Historical Quarterly* 45 (September 1944) ["got along tolerably well ... had to kill our work cattle" quote from p. 231]; **King**, Clarence, "The Falls of the Shoshone," *Overland Monthly* (October 1870) ["Intervals of light" quote from p. 383]; **Knight**, Oliver, "Robert E. Strahorn, Propagandist for the West," *Pacific Northwest Quarterly* 59 (January 1968) ["You can't put your finger" quote from p. 44]; **Limbaugh**, Ronald H., *Rocky Mountain Carpetbaggers: Idaho's Territorial Governors, 1983–1890* (1982); **Lovin**, Hugh T., "Water, Arid Land, and Visions of Advancement on the Snake River Plain," *Idaho Yesterdays* 35 (Spring 1991):3-19; **Madsen**, Brigham D., *The Shoshoni Frontier and the Bear River Massacre* (1985); **Maguire**, James H., ed., *The Literature of Idaho: An Anthology* (1986); **Mercier**, Laurie, "Women's Role in Montana Agriculture," *Montana* (Autumn 1988) ["economic linchpins" quote from p. 52]; **Mercier**, Laurie, and Carole Simon-Smolinski, eds., *Idaho's Ethnic Heritage*, 3 vols. (1990); **Merrill**, Irving R., ed., *Bound for Idaho: The 1864 Trail Journal of Julius Merrill* (1988) ["black vomit" quote from p. 101]; **Morgan**, Dale L., ed., *The Overland Diary of James A. Pritchard* (1960) ["the valley ... grass continues fine" quote from p. 103]; **Murphy**, Paul L., "Early Irrigation in the Boise Valley," *Pacific Northwest Quarterly* 44 (October 1935):177-84; **Nash**, John D., "The Salmon River Mission of 1855," *Idaho Yesterdays* 2 (Spring 1967):22-31; **Ourada**, Patricia, *Migrant Workers in*

Idaho (1990); **Palmer**, Joel, "Journal of Travels Over the Rocky Mountains to the Mouth of the Columbia River Made During the Years 1845 and 1846," *Early Western Travels, 1748–1846*, Reuben Gold, ed., vol. 30 (1904–1907) ["clear and beautiful

... abounds in fish" quote from p. 8]; **Palmer**, Tim, *The Snake River: Window to the West* (1991); **Paul**, Rodman W., "When Culture Came To Boise: Mary Hallock Foote In Idaho," *Idaho Historical Series* 19 (March 1977) ["I forsee the time" quote from p. 12]; **Powell**, William J., "I'll Take It If It's Legal," *The Pacific Northwesterner* 13 (Summer 1969):33-40, (Fall 1969):55-64; **Rich**, E.E., and A.M. Johnson, eds., *Peter Skene Ogden's Snake Country Journal, 1824–25 and 1825–26* (1950); **Robinson**, Michael C., *Water for the West: The Bureau of Reclamation, 1902–1977* (1979); **Rollins**, Philip A., ed., *The Discovery of the Oregon Trail: Robert*

Stuart's Narratives of his Overland Trip Eastward from Astoria in 1812–13 (1935) ["light green ... lovelier and wider," "constricted, full of rapids," "detested shrubs ... terrific appearance" quotes from pp. 290, 295, 114]; **Ross**, Alexander, *The Fur Hunters of the Far West*, Kenneth A. Spaulding, ed. (1956) ["altogether ... delightful" quote from p. 136]; **Schwantes**, Carlos A., *In Mountain Shadows: A History of Idaho* (1991); **Sharp**, Joe H., "Crossing the Plains in 1852," *Oregon Pioneer Association Transactions* (1985) ["Snake River ... inaccesible depths" quote from p. 93]; **Sharp**, *Shoshoni and Idaho Perspectives* (1989); **Strahorn**, Robert E., *The Resources and Attractions of Idaho Territory* (1990); **Valora**, Peter J., *A Historical Geography of Agriculture in the Upper Snake River Valley, Idaho* (Master's thesis, University of Colorado, 1986); **Wells**, Merle W., *Idaho, Gem of the Mountains* (1985); **Williams**, Glyndwr, ed., *Peter Skene Ogden's Snake Country Journals, 1827–28 and 1828–29* (1971); **Young**, F.G., ed., *The Correspondence and Journals of Captain Nathaniel J. Wyeth, 1831–36* (1899) ["strong volcanic," "fertile character" quotes from pp. 161, 164]; **Young**, Virgil M., *The Story of Idaho: Centennial Edition* (1989); **Yost**, George, *Idaho, The Fruitful Land* (1980).

CH. 6: POLITICAL LANDSCAPE

Andrus, Cecil, Gov., "high water mark" quote from a press release (March 28, 1993); **Arrandale**, Tom, "The Battle for Natural Resources," *Congressional Quarterly* (1984); **Beal**, Merrill D., and Merle W. Wells, *History of Idaho*, 3 vols. (1959) ["tamper with the headgate" quote from vol. 2, p. 122]; **Bowler**, Peter, interviewed by Stephen Steubner (March 3, 1993); **Brown**, Janice, "major economic impact" quote from public testimony, House Resources and Conservation Committee hearing on the Henry's Fork Basin Comprehensive Water Plan (February 12, 1993); Brown, interviewed by Stephen Steubner (April 18, 1993); **Chapman**, Sherl, interviewed by Stephen Steubner (February 18, April 12, 1993); **Christianson**, Steve, "so disgusted" quote from public testimony, joint House-Senate hearing on the Henry's Fork Basin Comprehensive Water Plan (February 21, 1992); **Corlett**, John, "Nowadays, Water Issues Breed Strange Bedfellows," *Idaho Statesman* (March 14, 1983); **Diehl**, Ted, interviewed by Stephen Steubner (April 16, 1993); **Egan**, Timothy,

"Ranchers vs. Rangers Over Land Use," *New York Times* (August 19, 1990) ["There's an old saying in the Forest Service," "he's going to have an accident" quotes from p. A20]; **Everhardt**, William, *The National Park Service* (1984); **Ford**, Pat, "The View from the Upper Basin," *Western Water Made Simple* (1987):86-96; **Foss**, Phillip, *Politics and Grass* (1960); **Gramer**, Rod, "Snake River Solutions—as Shortsighted as Before," *Idaho Statesman* (March 20, 1983); **Green**, Dean, guest commentary, *Post-Register* (April 14, 1992) ["vote against motherhood," "immeasurable value" quotes from p. A11]; **Hawkins**, Stan, interviewed by Stephen Steubner (November 16, 1992); **Idaho Army National Guard**, "Picture that task" quote from Natural Resources Conservation Award nomination (April 1993); **Idaho Statesman**, "Idaho Power Co. Offers Irrigators a Compromise," (March 17, 1983); **Idaho Statesman**, "Andrus Tours N-site, Declines Comment," (October 19, 1988); **Lewis**, C. S., *The Abolition of Man* (1943) ["Man's power over nature" quote from p. 35]; **Little**, Charles E., "The Challenge of Greater Yellowstone," *Wilderness* (Fall 1987) ["go beyond glittering generalities"

quote from p. 54]; **Loertscher**, Tom, "consumptive use of water" quote from House of Representatives debate, Idaho Legislature (March 29, 1992); **Loosli**, Lynn, "fished the Henry's Fork" quote from House of Representatives debate, Idaho Legislature (March 29, 1992); **Lyman,** Ann, ed., "Water Rights Landmark Case in Idaho," *Idaho's Water and Energy* (July 1980):20; **Madison**, James, *The Federalist Papers, No. 10* (1787); **Marston**, Betsy, "Flame and Blame," *High Country News* (November 7, 1988):10-19; **Mellen**, Jon, interviewed by Stephen Stuebner (May 18, 1992); **Nagel**, Joe, "An open sewer," *Post-Register* (March 21, 1993); **National Academy of Science**, *A Report to the Advisory Committee to the National National Park Service*

on Research* (1963); **Noh**, Laird, "Interim re-election protection" quote from Senate debate, Idaho Legislature (March 30, 1992); **Oakley**, Glenn, "Nuclear Fall Out: Weapons Factory Plans for INEL Spark Debate," *Focus* 9 (Summer 1989) ["nothing has left the site" quote from p. 32]; **Reed**, Scott W., "The Other Uses of Water," *Idaho Yesterdays* 30 (Spring/Summer 1986) ["state water plan" quote from p. 37]; Reed, "Fish Gotta Swim," *University of Idaho Law Review* 28 (1991-92) ["as a matter of legal analysis" quote from p. 662]; **Robinson**, Glenn, *The Forest Service* (1975); **Schneider**, Keith, "Idaho Out Front," *Post-Register* (March 11, 1990); **Seidman**, Harold, *Politics, Position, and Power* (1986) ["When conflicts result" quote from p.

223]; **Shaw**, Bill, "Lone Ranger Rides Again," *People* (October 2, 1990) ["worse than anyplace I've seen" quote from p. 49]; **Siddoway**, James, "can't build a tax base" quote from public testimony, joint House-Senate hearing on the Henry's Fork Basin Comprehensive Water Plan (February 21, 1992); **Stacy**, Susan, ed., *Conversations* (1990) ["use it or lose it" quote from pp. 94-5]; **Stapilus**, Randy, *Paradox Politics: People and Power in Idaho* (1988) ["special interests" quote from p. 49]; **Steenhof**, Karen, interviewed by Stephen Steubner (April 20, 1993); **Stegner**, Wallace, *Where the Bluebird Sings to the Lemonade Springs: Living and Writing in the West* (1992) ["Considering the mood" quote from p. 312]; **U.S. Bureau of Land Management**, Boise District, "proposed projects" quote from Boise District manager David Brunner in a BLM press release (November 16, 1989); BLM, Boise District, *Final Environmental Statement—Snake River Birds of Prey Conservation Area* (1979); BLM, Boise District, *Snake River Birds of Prey Management Plan* (August 1985); BLM, Boise District, *Final Environmental Assessment Land Report* (November 1988); BLM, Boise District, *Final Environmental Impact Statement for Agricultural Development for Southwest Idaho* (1979) ["increasing tendency" quote from p. 3]; BLM, Boise District, "Realignment of Mountain Home Air Force Base and Proposed Expanded Range Capability," *Saylor Creek Environmental Impact Statement* (1990); BLM, Boise District, Sagebrush Rebellion, Inc., "not major national resources" quote from a typewritten form letter entitled "People vs. Birds of Prey" (August 25, 1980); **U.S. Bureau of Land Management**, State of Idaho, *Bureau of Land Management in Idaho: 1989 Facts Book* (1990); **U.S. House of Representatives**, Committee on Interior and Insular Affairs, *Oversight Hearings on the Greater Yellowstone Ecosystem* (October 24, 1985); **U.S. Interior Dept.**, Subcommittee on Irrigation and Reclamation, "Field Hearing on Hells Canyon Project," Boise (April 4, 1955) ["Water is to Idaho," "We are seeking" quotes from pp. 607, 4]; **U.S. National Park Service**, Craters of the Moon National Monument, *Resource Management Plan* (1987); **University of Idaho**, College of Forestry, *A Review of Scientific Research* at *Craters of the Moon National Monument* (1988); **Wilkinson**, Charles F., *Crossing the Next Meridian Land, Water, and the Future of the West* (1992) ["Lords of Yesterday," "code of the West," "no right to a monopoly," "take charge of the future" quotes from pp. 88-9, 113, and 304]; **Worster**, Donald, *Under Western Skies: Nature and History in the American West* (1992) ["never really turn all of nature into artifact" quote from p. 254]; **Wyatt**, Margaret, *Snake River Birds of Prey Cultural Resource Management Plan* (1985).

EPILOGUE

Bradley, Carol, "Opponents Promise Valiant Effort to Kill Grazing Fee Hike," *Idaho Statesman* (June 20, 1991); **Brackett**, Bert, interviewed by Todd Shallat (December 30, 1993); **Carter**, David, interviewed by Todd Shallat (January 4 and 5, 1994); **Danes**, Ann, interviewed by Todd Shallat (January 5, 1994); **Day**, Ernie, interviewed by Todd Shallat (July 25, 1991); **Etlinger**, Charles, "A Look into the Future," *Idaho Statesman* (July 7, 1991); **Ford**, Pat, "View from the Upper Basin," *Western Water Made Simple*, Ed Marston, ed. (1987) [Larry Echohawk quote from p. 91]; **Foner**, Eric, and Jon Wiener, "Fighting for the West," *Nation* (July 29/August 5, 1991) [Howard Lamar quote, p. 163]; **Graham**, Bill, interviewed by Todd Shallat (June 25, 1991); **Grunder**, Scott, interviewed by Todd Shallat (December 29, 1993); **Gulick**, Bill, *Snake River Country* (1971); **Idaho Statesman**, "Sho-Bans Should Limit Salmon Catch to Ceremonial One," (July 3, 1991); **Mosley**, Jeffrey C. et al., *Seven Popular Myths About Livestock Grazing on Public Lands* (1990); **Reisner**, Marc, *Cadillac Desert: The American West and Its Disappearing Water* (1983); **Palmer**, Tim, *The Snake River: Window to the West* (1991); **Schellbach**, Ruth, interviewed by Todd Shallat (January 26, 1994); **Sojka**, Bob, interviewed by Todd Shallat (January 24, 1994); **Swan**, George, interviewed by Todd Shallat (December 30 and 31, 1993); **Wells**, Merle, "Idaho Indians" (typewritten, 1991); **Worster**, Donald, *Rivers of Empire: Water, Aridity, and the Growth of the American West* (1985) ["Oh

faith rewarded ... " Mormon hymn quote from p. 61].

PHOTO CREDITS

Front Matter: p. 1 Glenn Oakley; 2-3 Steve Bly; 4-5 Steve Bly; 6-7 Glenn Oakley #67040312; 8-9 Randy Kalisek/F-Stock Inc. cat. #29 53-32; 10-11 Idaho State Historical Society; 12-13 art by Gwenetta Bates.

Seeing the Land: pp. 14-15 Steve Bly; 15 Steve Bly (bottom); 16 chromolithograph by Julius Bien after a sketch by Gilbert Munger for the U.S. Geological Survey courtesy University of Idaho Library Special Collections (top), Glenn Oakley #67040317 (left), Steve Bly (right); 17 Idaho State Historical Society #80-94.38 (top), photo by Clarence E. Bisbee courtesy Boise State University Library Special Collections MSS #80-12025; 18 art by Gwenetta Bates (top), photo by Chuck Scheer courtesy Idaho Museum of Natural History — Idaho State University; 19 Idaho Museum of Natural History — Idaho State University #253 (detail, top left), etching by James David Smillie (1859) National Portrait Gallery — Smithsonian Institution #NPG.83.69 (detail, top right), National Anthropological Archives — Smithsonian Institution neg. #32549; 20 photo by Robert Limbert courtesy Boise State University Library Special Collections MSS #80-72 (top left), photo by Chuck Scheer from *Atlas of Idaho Territory, 1863–1890* courtesy Merle Wells and Idaho State Historical Society (detail, top right), Idaho State Historical Museum #1975.41.4; 21 U.S. Geological Survey Bulletin 199 courtesy University of Idaho Library Special Collections (top left), U.S. Geological Survey courtesy Idaho National Engineering Laboratory (top right); 22 Idaho State Historical Society #73-51.161/k (top left), photo by Chuck Scheer

courtesy Idaho State Historical Museum; 23 N. S. Nokkentved/*Twin Falls Times News*, art by Gwenetta Bates.

Volcanic Crescent: pp. 24-25 Steve Bly; 25 Chuck Scheer (bottom); 26-27 map by Richard Pike and Gail Thelin courtesy U.S. Geological Survey (detail); 26 Pike/Thelin U.S. Geological Survey map #I-2206 (detail); 28 Glenn Oakley; 29 art by Gwenetta Bates adapted from Time-Life Books series *Planet Earth: Volcano* (1982); 30 Hawaii Natural History Association courtesy U.S. Geological Survey; 31 Glenn Oakley (top left, center left, and bottom right), Dave Clark (top right), Bill Hackett (center right and bottom left); 32 art by Gwenetta Bates adapted from Time-Life Books series *Planet Earth: Volcano* (1982); 33 Pike/Thelin U.S. Geological Survey map #I-2206 (detail, top), Bill Bonnichsen; 34-35 Glenn Oakley; 36 Bill Bonnichsen (top), U.S. Geological Survey portraits #172 (detail); 37 Pike/Thelin U.S. Geological Survey map #I-2206 (detail, top), Bill Bonnichsen; 38 fig. 93 from G. P. Merrill's *First 100 Years of American Geology* (Hafner Pub., 1964) courtesy Idaho Geological Survey (top), Bill Bonnichsen; 39 Pike/Thelin U.S. Geological Survey map #I-2206 (detail); 40 art by Gwenetta Bates; 41 Glenn Oakley; 42-43 Glenn Oakley, 43 lithograph by Julius Bien after a photo by T. H. O'Sullivan for the U.S. Geological Survey courtesy University of Idaho Library Special Collections (top right); 44 Pike/Thelin U.S. Geological Survey map #I-2206 (detail, top), Bill Hackett; 45 from *Geological Society of America Memorials 18* (1988) courtesy Washington State University — Owen Science and Engineering Library; 46 Glenn Oakley (top left and bottom right), Bill Hackett (top right and bottom left); 47 Bill Hackett (top), Dave Clark; 48-49 Dave Clark; 50 Glenn Oakley; 51 Bill Hackett (top), Glenn Oakley; 52 Glenn Oakley; 53 Pike/Thelin U.S. Geological Survey map #I-2206 (detail); 54 Glenn Oakley; 55 Bill Hackett; 56 art by Gwenetta Bates adapted from Time-Life Books series *Planet Earth: Volcano* (1982); 57 Yellowstone National Park; 58 Bill Hackett (top), photo by W. H. Jackson courtesy U.S. Geological Survey #503 (center right), plate 66 from Hayden Survey 11th Annual Report (1877) courtesy University of Idaho Library Special Collections; 59 Glenn Oakley, art by Gwenetta Bates; 60-61 Glenn Oakley.

A Climate of Change: pp. 62-63 *Restoration of a Devonian Forest* by C. R. Knight courtesy Department of Library Services — American Museum of Natural History tr. #9923 (detail); 63 Michael Wickes/Wild World Productions #3922 (bottom); 64-65 maps by Chuck Scheer; 66 photo by Robert Limbert courtesy Boise State University Library Special Collections MSS #80-250; 67 art by Gwenetta Bates (top), Hagerman Fossil Beds National Monument; 68 photo by Chuck Scheer courtesy Idaho Museum of Natural History — Idaho State University; 69 Museum of Zoology/ Fish Division — University of Michigan courtesy Gerald R. Smith; 70-71 art by Gwenetta Bates adapted from a brochure from the Smithsonian exhibit The Emergence of Man; 71 Chuck Scheer (bottom); 72 Chuck Scheer (top right), Glenn Oakley; 73 Glenn Oakley; 74-75 photo by Jim Nay courtesy Grace Phillips

Johnson Gallery — Phillips University; 75 Glenn Oakley (bottom); 76 Idaho State Historical Society #66-4.46/b (top), Idaho State Historical Society #83-75.1; 77 Idaho State Historical Society MS #544 Box 7 Pam. 2; 78 Glenn Oakley (top), Chuck Scheer; 79 Glenn Oakley; 80 photo by Robert Limbert courtesy Boise State University Library Special Collections MSS #80 10945; 81 Idaho State Historical Society #73-51.23/a (top), Idaho Power Co.; 82 Glenn Oakley; 83 photo by Clarence E. Bisbee courtesy Idaho State Historical Society #73-221.755, art by Gwenetta Bates; 84 Monte La Orange/Idaho Falls Post Register; 85 Idaho Dept. of Fish and Game; 86 Dave Clark, Joel Muir (lichen, bottom right); 87 U.S.D.A. Forest Servive Intermountain Research Station courtesy Nancy Shaw; 88 art by Gwenetta Bates (top), photo by Robert Limbert courtesy Boise State University Library Special Collections MSS #80 10611; 89 Glenn Oakley #11060117; 90 photo by Michael Wickes/Wild World Productions #4054 (top), Dave Clark; 91 photo by Michael Wickes/Wild World Productions #47515; 92 Chuck Scheer (center left), Dave Clark; 93 William H. Mullins #60004-90504 (top left), Dave Clark (top center), Joel I. Mur; 94 William H. Mullins; 95 Glenn Oakley; 96 Tom Murphy/Wilderness Photography Expeditions; 97 art by Gwenetta Bates.

Native Traditions: pp. 98-99 Chuck Scheer; 99 photo by Chuck Scheer courtesy The Herrett Museum — College of Southern Idaho (bottom); 100-101 maps by Gwenetta Bates and Chuck Scheer; 101 photo by Chuck Scheer courtesy The Herrett Museum — College of Southern Idaho #TN B 86 1 (left); 102 George Herben/Alaska Stock Images #00400.098.02-06; 103 timeline photo by Chuck Scheer courtesy

Idaho Museum of Natural History — Idaho State University #10BN153-448 (bottom left), #10JE6 10260 (bottom right), #10JE6 10249 (center), #10JE6 10870 (top left), courtesy Ruth Gruhn (bottom left), courtesy Idaho Museum of Natural History — Idaho State University; 104-105 from Dinosaurs, Mammoths, and Cavemen: The Art of Charles R. Knight (1982) courtesy Rhoda Knight Kalt; 104-105 point art by Jim Woods; 106 photo by Chuck Scheer courtesy The Herrett Museum — College of Southern Idaho (atlatl #1989-1, dart #1989-3, foreshaft #1985-144), point art by Jim Woods; 107 photos by Chuck Scheer courtesy The Herrett Museum — College of Southern Idaho (top and #ID 77 37 center), art by Gwenetta Bates adapted from an illustration by Daniel Meatte (bottom left), art by Bill West; 108 art by Gwenetta Bates, photo by DeCost Smith (1904) courtesy National Museum of the American Indian neg. #22367; 109 art by Gwenetta Bates adapted from an illustration by Jim Woods (top), photo by Tom Green courtesy Idaho State Historical Society #100E60 (center), photo by Chuck Scheer courtesy Idaho Museum of Natural History — Idaho State University; 110 photo by Chuck Scheer courtesy The Herrett Museum — College of Southern Idaho #1989-125 (top), courtesy Idaho Museum of Natural History — Idaho State University (center left), photos by Lloyd Furniss courtesy Idaho Museum of Natural History — Idaho State University #14 (bottom right) and #25; 111 photo by Glenn Oakley (top left), map by Gwenetta Bates adapted from a graphic by Jim Woods, Idaho State Historical Society #77-69.4; 112 photos by Lloyd Furniss courtesy Idaho Museum of Natural History — Idaho State University #28 (top), #82 (center), #34; 113 point art by Jim Woods, art by Gwenetta Bates adapted from an illustration by Daniel Meatte; 114 art by Gwenetta Bates; 115 photos by Chuck Scheer courtesy The Herrett Museum — College of Southern Idaho #ID 77 25 (top left), #ID 77 23 (top right), National Anthropological

Archives—Smithsonian Institution neg. #1704-D (detail); 116-117 lithograph by H. Steinegger after a sketch by Edmond Green courtesy Idaho State Historical Society #1988-4.67; 117 photo by Glenn Oakley; 119 Fire in a Missouri Meadow and a Party of Sioux Indians by George Catlin courtesy The Fine Arts Museums of San Fransisco #1979.7.24 (top right, detail), photos by Chuck Scheer courtesy Idaho Museum of Natural History—Idaho State University #3260 (center), #5336; 120 Utah State Historical Society #970.8 P.7; 121 photos by Chuck Scheer courtesy The Herrett Museum — College of Southern Idaho #Rep 87 10 (bow), #Rep 78 7 and 1985-82 (arrows), #1992-14 (quiver), art by Gwenetta Bates.

Confronting the Desert: pp. 122-123 photo by Clarence E. Bisbee courtesy Idaho State Historical Society #73-221.506/b colorized by Phillip McClain; 123 photo by Chuck Scheer courtesy Idaho Museum of Natural History — Idaho State University #3278 (bottom); 124-125 maps by Gwenetta Bates and Chuck Scheer; 125 photos by Chuck Scheer courtesy Idaho State Historical Museum #941.21 (hatchet-pipe), #1986.30.2 (glove), #1984.125.74 (beads), #1962.56.15 (bludgeon); 126 lithograph after a photo by J. Buchtel (1860) courtesy Oregon Historical Society neg. #ORHI 10126; 127 photo by Chuck Scheer courtesy Idaho State Historical Museum #1957.37.0 (top), Missouri Historical Society Library & Archives #523 A; 128 Idaho State Historical Society #68-60.2 (top left), Colorado Historical Society #17.954 (top center), lithograph by Francis D'Avignon after a daguerreotype by Mathew Brady courtesy National Portrait Gallery—Smithsonian Institution, photo by Chuck Scheer courtesy Idaho State Historical Museum #848 (rifle), #63.241.157 (powder horn); 129 State Historical Society of Wisconsin #WHi(X3)41859; 130 Joslyn Art Museum, Omaha, Nebraska (top), photo by Chuck Scheer from Costume Reference 5, The Regency by Marion Sichel (1978); 131 photo by Chuck Scheer courtesy Idaho State Historical Museum # MS/25 (top right), Idaho State Historical Society #78-97.12 (bottom left), #78-97.14; 132 art by Gwenetta Bates (top), photo by Larry Burke (center), lithograph by James Ackerman after a sketch by J. Schutz courtesy Idaho State Historical Society #60-181.5; 133 National Anthropological Archives — Smithsonian Institution #1704-A-3; 134-135 The Butler Institute of American Art, Youngstown, Ohio; 135 photo by Chuck Scheer courtesy Idaho Historical Museum #1990.127.5; 136 map by Gwenetta Bates; 137 lithographs by E. Weber & Co. courtesy Idaho State Historical Society #1893-B (top left), #1254-D (top right), lithograph by Julius Bien after a photo by T. H. O'Sullivan courtesy University of Idaho Library Special Collections; 138 Church Archives, The Church of Jesus Christ of Latter-day Saints #P216 1; 139 Collection of Museum of Church History and Art, Salt Lake City; 140 Utah State Historical Society #921 P.1 (top), #970.8 P.2, Steve Bly (bottom right); 141 Idaho State Historical Society #1079-A; 142 Glenn Oakley; 143 The Huntington Library, San Marino, California #Hague Coll. box 60(5) (top left), Idaho State Historical Society #63-238.12 (top right), #MS 2/693 (bottom left, detail) #76-138.60; 144 Idaho State Historical Society #MS 544 Box 8, Pam. 1 (top left), #1086 (bottom left, detail), #162-A (detail); 145 Idaho State Historical Society #73-221.270; 146 Idaho State Historical Society *Commissioner of Immigration, Labor and Statistics Seventh Biennial Report 1911–12* (top left detail, and center right), Boise State University Library Special Collections #MSS 80 473; 147 Idaho State Historical Society #60-111.19 (top left), #71.26.3 (bottom left), 75-129.1/a; 148 photo by Henry R. Griffiths Jr. courtesy Idaho State Historical Society #65-166.35; 149 Idaho State Historical Society #MSS 544 Box 5, Pam. 1 (left, detail), map by Gwenetta Bates; 150 Idaho State Historical Society # MS 544 Box 5, Pam. 2; 151 Idaho State Historical Society #MS 544 Box 8, Pam. 6 (top), #MS 544 Box 8, Pam. 5 (center), #MS 544 Box 7, Pam. 16; 152 Highway Maps, Idaho State Historical Library; 153 Idaho State Historical Society #61-164.89; 154 photos by Chuck Scheer courtesy Jackie Fretwell (top left and right, details), Idaho State Historical Society #73-20.15/A (left); 154-155 Idaho State Historical Society #2592-5; 155 Denver Public Library Western History Dept. neg. #F34308 (top), Idaho State Historical Society #73-221.193; 156 photo by Larry Burke; 157 photo by Chuck Scheer courtesy Idaho State Historical Museum #1972.24.0 (top left), Idaho Power Co., map by Chuck Scheer; 158 photo by Chuck Scheer courtesy Laura Gipson, photo by Chuck Scheer courtesy Idaho State Historical Museum #1970.181, 1968.21.7, and 1991.17.16; 159 Glenn Oakley; 160 photo courtesy Basque Museum and Cultural Center; 161 photo by Chuck Scheer courtesy Bob Sims, art by Gwenetta Bates.

Political Landscape: pp. 162-163 *The Sentinel* by John Killmaster; 163 Glenn Oakley (bottom); 164-165 maps by Chuck Scheer; 166 Terri Davis (top), Chuck Scheer; 167 Glenn Oakley; 168 Glenn Oakley #41021001 (left), photo by Chuck Scheer courtesy Idaho State Historical Museum #1810 D; 169 John T. Isaacs/Isaacs Photography #41; 170-171 Steve Bly; 171 Chuck Scheer (top), Steve Bly; 172 Glenn Oakley; 174 Boise State Library Special Collections/Idaho Statesman Collection #79-3.229 (top), #76-3.530; 175 Glenn Oakley; 176-177 Glenn Oakley; 176 Pat Costello (top), John T. Isaacs/Isaacs Photography; 177 Chuck Scheer; 178-179 Glenn Oakley; 179 map by Chuck Scheer; 180 Dave Boehlke (top left), Dana Quinney/Idaho Army National Guard; 181 William H. Mullins #60004-90321; 182 Glenn Oakley; 183 courtesy Idaho Governor's Office

(top), Dana Quinney/Idaho Army National Guard; 184 Glenn Oakley (top), courtesy Mountain Home Air Force Base; 185 Glenn Oakley; 186 Idaho State Historical Society — Cattlemen Assoc. Vertical File; 187 Steve Bly; 188-189 Michael Wickes/Wild World Productions #2215; 188 courtesy Lynn Jacobs; 189 *Chisolm Trail* courtesy Scotts Bluff National Monument; 190-191 Steve Bly; 191 Plummer/F-Stock; 192 Glenn Oakley #41030410 (top), Tom Ansbach; 193 photo by Chuck Scheer courtesy Idaho State Historical Society #1965.143.9, #58-95, #1986.53.1; 194-195 Glenn Oakley; 194 Kevin Clark/*Idaho Statesman*; 195 Glenn Oakley (bottom left), photo by Chuck Scheer courtesy Helen Birge; 196 Steve Bly ; 197 Glenn Oakley; 198 map by Chuck Scheer; 199 Idaho State Historical Society #MS 544 Box 2, Pam. 11; 200 Glenn Oakley #11080401 (top), Idaho State Historical Society #MS 544 Box 2, Pam. 11 (detail); 201 Sharon Chuba, Glenn Oakley (inset); 202 Glenn Oakley #67011414 (top), #35230219; 203 Glenn Oakley (top left), Chuck Scheer courtesy Dave Clark (top right), Henry H. Holdsworth/F-Stock; 204 Steve Bly; 205 Glenn Oakley #090185-12; 206 Peter Wollheim (top), Steve Bly; 207 Brent Smith, art by Gwenetta Bates.

Epilogue: pp. 208-209 Steve Bly; 209 courtesy Mrs. Robert W. Addison; 210-211 Glenn Oakley; 212 Steve Bly; 213 courtesy Micron Technology Inc.; 214 art by Gwenetta Bates; 214-215 Glenn Oakley.

"Two-Gun" Bob: Photos by Robert Limbert courtesy Boise State University Library Special Collections MMS #80; p. 216 #12030; 217 #10764; 218 #1311; 219 #1316 (top), #1759; 220-221 #199; 222 #10507 (top), #503 (center), #1025; 223 #252; 224 #484 (top), #10616; 225 #254; 226 #474; 227 #1315.

CAPSULE AUTHORS

E. B. Bentley: "Tools of the Trade."

Peter Bowler: "Hagerman Fossil Beds National Monument," "Catastrophic Bonneville Flood," "Mollusks of the Middle Snake River," "White Pelican" and "Ord's Kangaroo Rat."

Dave Clark: "When the Desert Blooms" and "Trout Farming."

Bill Hackett: "Volcanic Primer" and "Craters of the Moon."

John Freemuth: "A National Park in Idaho?"

Josie Fretwell: "Preparing the Food" and "'Dreamers We Are.'"

Sandy Marostica: "Idaho's Sheep King."

Glenn Oakley: "Branches of the Snake" and "Snake River Sturgeon."

Mark Plew: "Ruth Gruhn," "Fishing" and "Earl H. Swanson."

Todd Shallat: "Clarence R. King," "Edward D. Cope," "Harold T. Stearns," "Ferdinand V. Hayden," "Don Crabtree," "Horse," "Names on the Land," "Early Publicists Market the Plain," "Wranglers and Turkey Tenders" and "'Two-Gun' Bob: Idaho's Showman."

Steve Stuebner: "Idaho's Famous Potatoes," "North Side Canal Co." "Idaho National Engineering Laboratory" and "Snake River Sho-Bans."

CONTRIBUTORS

E. B. BENTLEY specializes in the Pleistocene environments of the Great Basin. He received a Ph.D. in physical geography/geomorphology from the University of Oregon and is now a professor of geosciences at Boise State University.

BILL BONNICHSEN was the first to recognize the enormous Bruneau-Jarbidge caldera complex. He also discovered the large rhyolite lava flows that occurred in southwest Idaho and has described the explosive basaltic volcanoes that formed where Lake Idaho once stood. He grew up in the southern Idaho town of Filer, received his B.S. in geological engineering from the University of Idaho in 1960, attained his Ph.D. in geology from the University of Minnesota in 1968 and joined the Idaho Geological Survey in 1976.

JOHN FREEMUTH, an associate professor of political science and public administration at Boise State University, received his Ph.D. from Colorado State University in 1976. He is the author of *Islands Under Siege: National Parks and the Politics of External Threats* (University of Kansas, 1991), as well as numerous articles on public lands policy. He spends much of his time working with federal land agencies on various projects, and is currently at work on a book on the politics of ecosystem management for Kansas Press.

BILL HACKETT received his Ph.D. in geology from Victoria University in New Zealand in 1985. A specialist in the study of volcanoes and their deposits, he has done extensive research in New Zealand, Japan, Hawaii and the American West. His work on volcanism in Idaho has appeared in many books and journals, including Guidebook to the Geology of Central and Southern Idaho which was published by the Idaho Geological Survey in 1988. A former geology professor at Idaho State University, he has also been a scientist for the Idaho National Engineering Laboratory and is currently a geological consultant based in Pocatello.

GLENN OAKLEY has a B.A. in journalism from the University of Montana and studied in the graduate photography program at Ohio University. He works part time as a writer for *FOCUS* magazine at Boise State University and runs a photography business, specializing in outdoor and environmental subjects. His writing and photographs have appeared in a variety of national magazines and books.

MARK G. PLEW is chairman of the Anthropology Department at Boise State University. A Ph.D. from Indiana University, he has conducted archaeological research throughout the United States, Australia and South America and has published widely on various aspects of archaeology. A specialist in the recent prehistory of Idaho, he is the editor of *The Idaho Archaeologist* and director of the Boise State Archaeological Field School Program.

F. ROSS PETERSON is a native of Montpelier, Idaho. He received a Ph.D. from Washington State University in 1968. Since 1971 Peterson has taught history at Utah State University and currently directs the Mountain West Center for Regional Studies. He is the author of *Prophet Without Honor: Glen H. Taylor and the Fight for American Liberalism* (1974), *Idaho: A Bicentennial History* (1977) and other books and articles on Idaho and the West. He is currently writing a biography of former Idaho Sen. Frank Church.

TODD SHALLAT directs the public history program at Boise State University. A Ph.D. in applied history from Carnegie-Mellon University, Shallat specializes in the history of science, engineering and the environment. His articles on water and engineering have appeared in *Technology and Culture*, *Natural Resources Journal* and other British and American publications. *Structures in the Stream*, his fifth book, is forthcoming from the University of Texas Press.

STEVE STUEBNER is a professional free-lance journalist based in Boise. A graduate of the University of Montana, he has more than a decade of experience covering natural resource issues for the *Idaho Statesman*, the Portland *Oregonian*, *High Country News* and other national and regional publications.

INDEX